ORIGINS

ORIGINS

The Darwin College Lectures

Edited by

A.C. Fabian

Royal Society Research Professor
in Astronomy
University of Cambridge

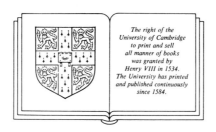

The right of the
University of Cambridge
to print and sell
all manner of books
was granted by
Henry VIII in 1534.
The University has printed
and published continuously
since 1584.

CAMBRIDGE UNIVERSITY PRESS

Cambridge
New York   New Rochelle   Melbourne   Sydney

Published by the Press Syndicate of the University of Cambridge
The Pitt Building, Trumpington Street, Cambridge CB2 1RP
32 East 57th Street, New York, NY 10022, USA
10 Stamford Road, Oakleigh, Melbourne 3166, Australia

First published in 1988

Printed in Great Britain at the
University Press, Cambridge

British Library Cataloguing in Publication Data
Origins
1. Universe. Origins 2. Man. Evolution
I. Fabian, A.C. II. Darwin College
523.1'2

Library of Congress Cataloguing in Publication Data
Origins: the Darwin College lectures/edited by A.C. Fabian.
p. cm.
Lectures delivered during the Lent Term of 1986.
Includes index.
Contents: Introduction/D. Hugh Mellor
Origin of the universe/Martin J. Rees
Origin of the solar system/David W. Hughes
Origins of complexity/Ilya Prigogine
Human origins and evolution/David Pilbeam
Origin of social behaviour/John Maynard Smith
Origins of society/Ernest Gellner
Origins of language/John Lyons.
1. Cosmology. 2. Human evolution.
3. Social evolution.
4. Language and languages – Origin.
I. Fabian, A. C., 1948-
II. Title: Darwin College lectures.
QB981.O75 1988
523.1-dc19   88-23435   CIP

ISBN 0 521 35189 8

Contents

Contributors

D. H. Mellor
Vice-master of Darwin College, and Professor of
Philosophy in the University of Cambridge

Martin J. Rees
Plumian Professor of Astronomy in the University of
Cambridge

David W. Hughes
Senior Lecturer in the Department of Physics in the
University of Sheffield

Ilya Prigogine
Professor of Physical Chemistry in the Free University of
Brussels and Director of the Center for Statistical
Mechanics and Thermodynamics in the University of
Texas at Austin

David Pilbeam
Professor of Anthropology in the Peabody Museum of
Harvard University

John Maynard Smith
Professor of Biology in the University of Sussex

Ernest Gellner
Professor of Social Anthropology in the University of
Cambridge

John Lyons
Master of Trinity Hall, Cambridge

Preface

University research covers a great range of subjects. To try to comprehend all of them would be foolish: life is too short, and anyway no one is good at everything. But most subjects are to some extent spectator sports. You needn't be a musician to appreciate some modern music – though no doubt it helps – nor a cosmologist to appreciate some modern cosmology. And many spectators have common interests in very different subjects: an interest in their origins, for example. A natural narcissism interests us in our own origins and, by a natural extension, in the origins of the very diverse parts and aspects of the world on which we depend.

There is, therefore, a predictable demand for series of public lectures by leading authorities in interdisciplinary topics such as origins. And not only for lectures: such interests are not confined to Cambridge, nor to one year. So the lectures might profitably be reprinted in a book, to make them available permanently to a wider audience.

But who is to arrange all this? In the nature of the case, it is not the business of any one university department. And although colleges cut across departmental boundaries, the main business of an undergraduate college is still to teach its students their own subjects. But the main business of a graduate college such as Darwin College is research, and especially aspects of it that relate, or are common to, different subjects. Arranging lecture series and books on such topics is a natural activity for a graduate college. Hence the sequence of lecture series and books of which this is the first.

The first Cambridge college to undertake such a project, we at Darwin have been encouraged in part by the example of our counterpart in Oxford, Wolfson College, which has organized several similar series of lectures and books. And in 1982 we organized the Darwin Centenary Conference and arranged the publication of its proceedings by Cambridge University Press under the title *Evolution from Molecules to Man*. The success of that conference and that book has been the main stimulus to our present enterprise. And it seemed especially apt for a College named after Charles Darwin to follow his Centenary Conference, and start our sequence, with a series of lectures and a book on origins. So, too, it clearly seemed to the massive audiences for these lectures, who throughout the Lent Term of 1986 packed the University's largest lecture theatre week after week to hear them. The demand in fact exceeded all our predictions: and those who were there know how well our lecturers met it. Just how well, readers may now judge for themselves; and we hope they will be encouraged thereby to support with equal enthusiasm the next, equally, apt, 1987 series of Darwin College Lectures, on *Man and the Environment*.

D. H. Mellor

Introduction

In discussing origins, it makes more than conventional sense to begin at the beginning, since the origin of everything is also the origin of all other origins. Hence our first chapter by Martin Rees, which sets us off with a bang: the hot big bang that was probably the origin of our expanding Universe. Not that modern cosmology starts with the big bang. Rather, it works back to it, using our knowledge of gravity and nuclear physics to trace and explain the formation, history and workings of galaxies and stars, including the creation of the matter we are made of. We may never actually reach the big bang, because the nearer we get, the less we know and the more we can only speculate. Still, Rees gets us close, showing how informed speculation can now reach back to the first 10^{-36} or even 10^{-43} seconds. And if we still want the big bang explained, Rees reminds us that cosmology will not explain it anyway, since origins explain what follows them ('things are as they are because they were as they were'), not vice versa.

In chapter 2 we come closer to home, with the origin of the Solar System. This is harder to settle than the origins of stars and galaxies, because we have only one visible specimen to study. David Hughes takes us through the facts and a range of theories. The more or less known facts concern the similar ages of Earth and Sun, and the latter's evolution; the masses and sizes of the planets, and hence their densities and probable composition; their orbits and hence their angular momenta; the prevalence of planets among nearby stars; meteorites; and possible ways of forming planets. The theories include: a spinning cloud flinging off rings that condense into planets, the residue condensing into the Sun; a passing star pulling the matter of the planets from the Sun; the Sun picking it up from interstellar dust and gas. Hughes' favourite is that Sun and planets condensed from a single cloud, a mechanism that would give planets to about one star in five – some 20 billion in our galaxy.

These developments, of the Universe and of the Solar System, involve irreversible change and increasing complexity. In chapter 3 Ilya Prigogine asks how complexity can arise and increase. He ascribes it to instability in dynamical systems, which can make even deterministic ones – e.g. the orbits of comets – unpredictable and irreversible. He also finds these effects in the non-equilibrium thermodynamics of chemical and other processes, where the interplay of a few variables may 'attract' a system like the atmosphere to many stable states, but with much fluctuation. Prigogine finally speculates that such irreversible processes might have produced the whole material Universe by the bifurcation of a primordial vacuum into matter and gravity.

Chapter 4 brings us to human life in particular. When and how we evolved depends on what it took to become human: a complex question. David Pilbeam takes major stages to be: bipedal hominids appearing about 5 million years (my) ago; their diversifying between 5 and 1.5 my ago; the use of stone-flakes, greater brain size and more meat-eating in *Homo* over 2 my ago; a non-ape-like omnivorous scavenging and/or hunting phase with stone tools and probably speech; modern humans and perhaps modern language capacity by 40 or more millennia ago; and a shift in the last 10 millennia from gathering food to producing it. He sketches a range of current theories about when, why, how and by what steps these changes occurred; the relevant fossil, genetic, behavioural and other evidence; and how much it may ultimately tell us about our ancestry.

After matter and life there comes the still greater complexity of society – animal and human. In chapter 5 John Maynard Smith asks how the co-operative behaviour which animal societies require could evolve. He answers that co-operation is often synergistic and thus evolutionarily stable, and gets going because animal societies are composed of relatives, who will spread 'co-operative' genes. Co-operation with non-relatives can also spread by reciprocal altruism on a tit-for-tat basis. Human society, being self-conscious, needn't evolve genetically; but our capacity for it, Maynard Smith argues, probably did, because of its synergistic effects on evolutionary fitness. Finally, Maynard Smith considers in a range of cases how far loyalty to a human group may make it, or depend on its being, a breeding group.

In chapter 6 Ernest Gellner considers specifically the origins of our own societies. He starts from their diversity. It isn't genetic: children can be socialized into any society. The diversity of our societies thus implies that each one severely – and differently – restricts our possible behaviour. How is that done? Ritual, says Gellner; but ritual alone won't explain our contractual and coercive social structure. The origin of that, he thinks, lies in our starting to produce and store food and so needing to distribute and defend it. That in turn needed language and then writing: hence the production, storage and distribution of concepts and doctrines as well as food. But that allows people to envisage new ways of behaving: it strengthens logical but weakens social coherence. So to preserve society we then needed a shift from ritual to doctrine: gods to enforce, but also to be bound by, systems of social concepts. But then the concepts became debatable, if only among their clerical articulators, and our loyalty eventually gets transferred from specific doctrines to our modern social systems for deciding and enforcing them.

All this, of course, demands language, perhaps our most distinctive and important social trait, whose origins John Lyons looks at in our last chapter. He distinguishes the origins of particular languages from the origins of language itself. Knowing the former is no help with the latter, because earlier languages, and those of so-called primitive people, are no simpler or less highly evolved than ours. Nor does Lyons think in the case of language that 'ontogeny recapitulates phylogeny', though neither is yet understood. But in both, he distinguishes language from speech. He thinks human language was probably gestural in origin, like the rudimentary languages taught to chimpanzees, which compare in complexity with the 'telegraphic' utterances of two-year-old children. But the grammatical and descriptive complexity of adult language, which *is* associated with speech, requires capacities beyond those of other terrestrial species, capacities we may have acquired in only the last 40–100 millennia.

All our lecturers emphasize how far we are from knowing all about our origins. But between them they also show clearly how far we have come already, and where we now stand. The lectures in this book, we venture to hope, will provide as good an origin for new work on origins as they already have for our new sequence of Darwin College lecture series.

D. H. Mellor

[1]

Origin of the Universe

Martin J. Rees

'Whilst this planet has gone cycling on according to the fixed law of gravity, from so simple a beginning, endless forms most . . . wonderful have been, and are being evolved.' These concluding words from *The Origin of Species* could serve as a text for later lectures in this series. But I shall be trying to describe what happened *before* the era that Darwin took as his 'simple beginning': to set the Earth in a broader evolutionary context, and trace the origins of its constituent material back to the formation of our Galaxy – right back, indeed, to the first seconds of the so-called 'big bang' that initiated our expanding Universe.

Cosmologists cannot yet offer more than the rudiments of this overall cosmogonic scheme; but what should really surprise us is that there has been any progress at all. 'The most incomprehensible thing about the Universe is that it is comprehensible' is one of Einstein's best-known sayings. It expresses his wonder that the physical laws, which our brains are somehow attuned to make sense of, apparently apply not just in the lab but in the remotest parts of the Universe. This unity and interrelatedness of the physical world must impress all who ponder it. Later on, I shall venture towards some speculative fringes of the subject, but let us start with something quite well understood, the life-cycle of a star like our own Sun.

The Sun and stars

The Sun started life by condensing gravitationally from an interstellar cloud. It continued to contract until its centre became hot enough to ignite nuclear reactions. Gradual conversion of hydrogen into helium then releases enough energy to keep the Sun burning steadily, as a gravitationally confined fusion reactor, for about ten thousand million (ten billion) years. It has been shining for four-and-a-half billion years, and about five billion years from now the hydrogen in its core will run out. It will then swell up to become a red giant, engulfing the inner planets, before settling down to a quiet demise as a white dwarf.

The study of stellar evolution made little progress until the 1930s. Before that

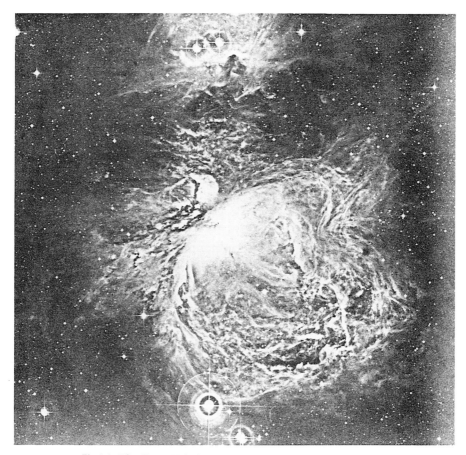

Fig.1.1. The Orion Nebula: a region where stars are now forming.

time, the physics of nuclear reactions was not understood. Indeed, the Sun's age was something that Darwin worried about. Lord Kelvin argued that gravitation must be the prime energy source, and that the Sun must inexorably contract as it loses heat. He calculated that unless sources now unknown to us are 'prepared in the great storehouse of creation', the Sun could last only 20 million years – a period more than ten times shorter than Darwin and the geologists thought comfortable. Only much later did laboratory physics reveal the nuclear fuel that Kelvin could not conceive of.

Stars are so long lived compared with astronomers that we only have, in effect, a single 'snapshot' of each one. But the fact that there are so many of them makes up for this; and we *can* check our theories, just as we could infer the life-cycle of a tree by one day's observation of a forest. Of special interest are places like the Orion Nebula (Figure 1.1), where even now stars, perhaps with new solar systems, are condensing from glowing gas clouds; and star clusters (Figure 1.2), containing stars of different sizes which are thought to have formed at the same time.

Not everything in the cosmos happens slowly. Stars heavier than the Sun

Fig.1.2. A globular star cluster: a self-gravitating system of several
hundred thousand coeval stars.

evolve faster, and some expire violently as supernovae. The best-known
instance is the Crab Nebula, the expanding debris of a stellar explosion seen
and recorded by oriental astronomers in AD 1054. In July of that year, Yang Wei
Te, the Chinese 'chief calendrical computer' (the counterpart of our Astronomer Royal, presumably), reported to the Emperor that a 'guest star' had
appeared. This star faded after a few months, leaving a remnant behind (Figure
1.3). Supernova explosions signify the violent end-point of stellar evolution,
when a star too massive to become a white dwarf exhausts its available nuclear

[3]

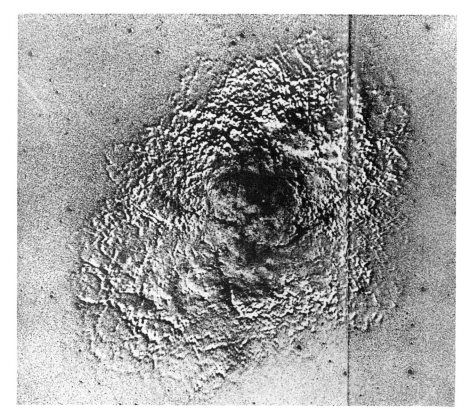

Fig.1.3. The Crab Nebula. This picture shows a positive print of the nebula superposed on a negative taken 15 years earlier. The 'bas relief' appearance of the filaments is graphic evidence that the nebula, which lies at a distance of about 5000 light years, has been expanding (at speeds of order 1000 km s^{-1}) since the supernova event in AD 1054.

energy. The star then faces an energy crisis. Its core catastrophically implodes, releasing so much gravitational energy that the outer layers are blown off. The centre of the star collapses to form a spinning neutron star (pulsar) only about 10 km across.

Supernovae may seem remote and irrelevant to our own origins. But on the contrary, only by studying the births of stars, and the explosive way they die, can we tackle such an everyday question as where the atoms we are made of came from. The respective abundances of the elements of the periodic table can be measured in the Solar System, and inferred spectroscopically in stars and nebulae. The proportions in which the elements occur display regularities from place to place which certainly demand some explanation (Figure 1.4).

Complex chemical elements are an inevitable by-product of the nuclear reactions that provide the power in ordinary stars. A massive star develops a kind of onion-skin structure, where the inner hotter shells are 'cooked' further up the periodic table (Figure 1.5). The final explosion ejects most of this

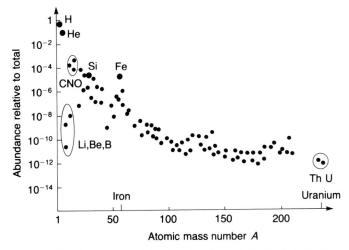

Fig.1.4. The abundances of the chemical elements in the Solar System are here plotted (on a logarithmic scale) as a function of atomic number. Note that only 2 per cent of all the matter is in elements heavier than hydrogen and helium. The relative abundances of the heavy elements are fairly standardized throughout the galaxy; in some of the oldest stars, however, these elements are all depleted relative to hydrogen and helium compared to abundance measurements in the Solar System (i.e. much less than 2 per cent of the total). No objects have been found to contain less than 23–25 per cent (by mass) of helium. It is believed that the material emerging from the 'big bang' was essentially just hydrogen and helium, and that heavier elements were synthesized in stars over the lifetime of the galaxy.

processed material. All the carbon, nitrogen, oxygen and iron on the Earth could have been manufactured in stars that exhausted their fuel supply and exploded before the Sun formed. The Solar System would then have condensed from gas contaminated by the debris ejected from earlier generations of stars. These processes of cosmic nucleogenesis can account for the observed proportions of different elements – why oxygen is common but gold and uranium are rare – and how they came to be in our Solar System.

Each atom on Earth can be traced back to stars that died before the Solar System formed. A carbon atom, forged in the core of a massive star and ejected when this exploded as a supernova, may spend hundreds of millions of years wandering in interstellar space. It may then find itself in a dense cloud which contracts into a new generation of stars. It could then be once again in a stellar interior, where it is transmuted into a still heavier element. Alternatively, it may find itself out on the boundary of a new solar system in a planet, and maybe eventually in a human cell. We are literally made of the ashes of long-dead stars.

This concept of *stellar nucleosynthesis*, due primarily to Hoyle, Fowler and Cameron, is one of the triumphs of astrophysics in the last 40 years. It sets our Solar System in a kind of ecological scheme involving the entire Milky Way galaxy. The particular mix of elements that we find around us is not *ad hoc*, but

[5]

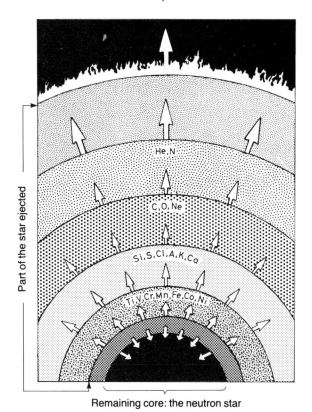

Part of the star ejected

Remaining core: the neutron star

Fig.1.5. The structure of a massive star before the final supernova outburst. The hotter inner shells have been processed further up the periodic table; this releases progressively more energy until the material is converted into iron ($A = 56$), the most tightly bound nucleus. Endothermic nuclear reactions occurring behind the shock wave that blows off the star's outer layers (explosive nucleosynthesis) can synthesize small quantities of elements beyond the 'iron peak' (see Figure 1.4).

the outcome of transmutation and recycling processes, whose starting-point is a young galaxy containing only the lightest elements. One thing this scheme could not explain, however, was helium – why this makes up so much of the mass of all stars, young or old. We shall come back to that later.

Galaxies and their active nuclei

Let us now extend our horizons to the extragalactic realm. It has been clear since the 1920s that our Milky Way, with its 10^{11} stars and scale of about a hundred thousand light years, is just one galaxy, similar to millions of others visible with large telescopes.

Galaxies are held in equilibrium by a balance between gravity, which tends to draw the stars together, and the countervailing effect of the stellar motions,

Fig.1.6. A disc galaxy, viewed at an oblique angle.

which if gravity did not act would cause the galaxy to fly apart. In some galaxies, our own among them, stars move in nearly circular orbit in giant discs (Figure 1.6). In others, the less photogenic ellipticals (Figure 1.7), the stars are swarming around in more random directions, each feeling the gravitational pull of all the others.

Our understanding of galactic morphology is tentative, maybe at the same

[7]

Fig.1.7. An elliptical galaxy.

Fig 1.8. The Sombrero galaxy – displaying a hybrid of disc and bulge morphology.

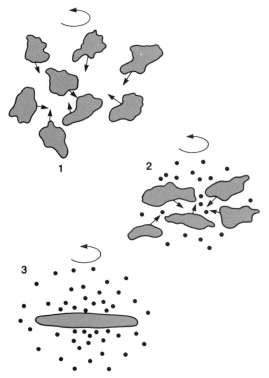

Fig.1.9. Schematic diagram showing three stages in the collapse of a protogalactic cloud with initial mass 10^{11}–$10^{12}\,M_\odot$ and radius exceeding 10^5 light years. This cloud would have irregular substructure (depicted here as a number of 'subclouds'). The subclouds merge and dissipate energy when they collide; stars, on the other hand, are most unlikely to collide. Stars that condensed out during the infall would retain complex three-dimensional orbits; but those that formed later, from gas that had already settled into a disc, would have more 'ordered' near-circular orbits predominantly in the plane of the disc. The resultant morphology (elliptical or disc-like) then depends on whether the timescale for star formation is shorter or longer than the collapse time. (There is, however, an alternative scheme for forming at least some elliptical galaxies. It could be that most galaxies form as discs, and that ellipticals result from *mergers* of disc systems that occur at a later stage.)

level as the theory of individual stars was 50 years ago: there is much boisterous debate and several competing theories, but little consensus on the details. There is, however, a widely adopted scenario that accounts qualitatively for the two basic types – discs and ellipticals (Figure 1.8). Let us suppose that all galaxies started their lives as huge turbulent gas clouds contracting under their own gravitation, and gradually fragmenting into stars (as depicted in Figure 1.9). The collapse of such a gas cloud is highly dissipative, in the sense that any two globules that collide will radiate their relative energy via shock waves, and will merge. The end result of the collapse of a rotating gas cloud will be a disc. This is the lowest energy state that such a cloud can reach if it loses energy but not its angular momentum. *Stars*, on the other hand, do not collide with each other, and cannot dissipate energy in the same fashion as gas clouds. This suggests that the *rate of conversion of gas into stars* is the crucial feature determining the type of galaxy that results. Ellipticals will be those in which the conversion is fast, so that most stars have already formed before the gas has had time to settle down in a disc. Contariwise, the disc galaxies will be those in which most star formation is delayed until the gas has already settled into a disc. The origin of these giant gas clouds is a cosmological question. But, given such clouds, the physics that determines galactic morphology is nothing more exotic or highbrow than Newtonian gravity and gas dynamics. This does not, of course, make the phenomena easy to quantify, any more than weather prediction is easy.

Some peculiar galaxies, though, are more than just a 'pile' of stars, and harbour intense superstellar activity in their centres. Nearby galaxies such as Centaurus A show this phenomenon in a mild way (Figure 1.10). But most extreme are the quasars, where a small region no bigger than the Solar System outshines the entire surrounding galaxy, and the so-called radio galaxies, whose most conspicuous output is not visible light, but radio waves. In such galaxies, where the central power exceeds a million supernova explosions in unison, gas and stars have accumulated in the centre, until gravity overwhelms all other forces and a *black hole* forms. Here we need more highbrow physics, Einstein's general relativity (the theory that 'matter tells space how to curve, and space tells matter how to move'). The genesis of this theory was unusual: Einstein was not motivated by any observational enigma but, rather, by the quest for simplicity and unity; it was invented almost prematurely in 1915, when any prospects of observing strong field gravity seemed remote. But ever

Fig.1.10. Centaurus A: a relatively nearby galaxy with an active nucleus, which is a radio source.

since active galaxies were discovered, relativists have been, in T. Gold's words 'not merely magnificent cultural ornaments, but actually relevant to astrophysics'.

The renaissance in general relativity in the 1960s stemmed not only from observational advances, but also from the deployment (by Penrose, Hawking, and others) of novel mathematical techniques. Gravitational collapse, however asymmetrically it occurred, was found to lead to black holes whose properties could be exactly specified in terms of just two parameters: mass and spin. To quote S. Chandrasekhar: 'The black holes of nature are the most perfect macroscopic objects there are in the Universe: the only elements in their construction are our concepts of space and time. And since the general theory of relativity provides only a single family of solutions for their descriptions, they are the simplest objects as well.'

Black holes have now entered the general vocabulary, if not yet the common understanding. They are objects whose gravitational field is so strong – where space is so strongly curved – that not even light can escape. Black holes are the 'ghosts' of dead stars or galaxies – objects that have collapsed, cutting themselves off from the rest of the Universe, but leaving a gravitational imprint frozen in the space they have left. To physicists, gravitational collapse is important because the central 'singularity' must be a region where the laws of classical gravitation are transcended, and one needs some unified physical theory to understand what is going on. Black holes bear on our general concepts of space and time, because near them space behaves in peculiar ways that are highly non-intuitive to us. For instance, time would stand still for an observer who managed to hover or orbit just outside a hole's surface, and he could witness the entire future of the external Universe in what to him seemed quite a short period. Even stranger and less predictable things might happen if one ventured inside.

Most theorists believe that the central prime mover in active galaxies involves a spinning black hole as massive as a hundred million suns, fuelled by capturing gas, or even entire stars. This captured debris swirls downward into the hole, carrying magnetic fields with it and moving nearly at the speed of light. At least 10 per cent of the rest mass energy of the infalling material can be radiated; further energy can be extracted from the hole's spin. Some of us are hopeful that these ideas can be put on a firm quantitative basis, just as our theories of stellar evolution have been. If so, this would offer real opportunities to probe the properties of strong gravitational fields where relativistic effects could be crucial.

Black holes were, in essence, conjectured more than 200 years ago. John Michell, an under-appreciated polymath of eighteenth-century science, published a paper in *Philosophical Transactions of the Royal Society* in 1784. He noted that the escape velocity from the Sun is about $\frac{1}{500}$th the speed of light, but would be 500 times larger for an object having the same density as the Sun but 500 times its radius. He noted that light would be made to return towards such an object 'by its own proper gravity'. Laplace made the same point a decade later, but if we accord Michell his deserved priority, 1984 was the black hole

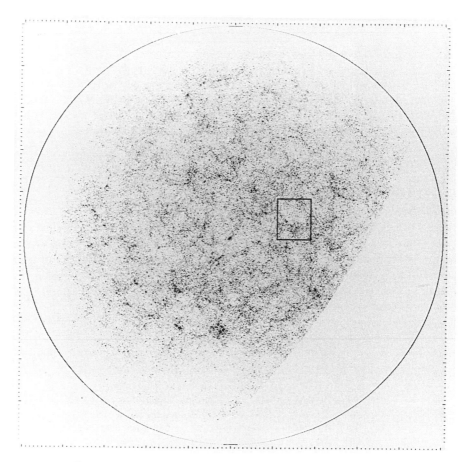

Fig.1.11. The distribution over the sky of the brightest million galaxies. Although some clustering is apparent, there are roughly equal numbers in all areas of sky surveyed. The larger the scale, the smoother the Universe appears.

bicentennial. Of course, Michell's argument used the ballistic theory of light and Newtonian gravity, not the relativity that is really needed.

The expanding Universe

The other arena where Einstein's theory is crucial is cosmology, the description of our Universe as a single dynamical entity. Scientific cosmology is the study of a unique object and a unique event. No physicist would happily base a theory on a single unrepeatable experiment. No biologist would formulate general ideas on animal behaviour after observing just one rat, which might have psychoses peculiar to itself. But we plainly cannot check our cosmological ideas by applying them to other universes. Nor can we repeat or rerun the past, though the finite speed of light allows us to sample the past by looking at very remote objects.

Despite having these things stacked against it, scientific cosmology has proved possible, but only because the observed Universe, in its large-scale structure, is simpler than we had any right to expect. It is of course sensible methodology to start by making simplifying assumptions about homogeneity, symmetry, etc., and cosmologists did this: indeed, back in the 1920s, Friedman, Lemaître, and others devised cosmological models based on Einstein's relativity. But what *is* surprising is that these models remain relevant, and the simplifying assumptions have been vindicated.

We are talking now about vast intergalactic scales of distances. With a large telescope, one can probe billions of light years into space. To the cosmologist, even entire galaxies are just markers or test particles scattered through space, which indicate how the material context of the Universe is distributed.

Galaxies are clustered. Some are in small groups such as our own Local Group, of which the Milky Way and the Andromeda galaxy are the dominant members. Others are in big clusters with hundreds of members. But on the really large scale, the Universe genuinely does seem simple and smooth. If one imagined a box whose sides were one hundred million light years (dimensions still small compared to the observable Universe) its contents would look about the same wherever we placed it. In other words there is a well-defined sense in which the observable Universe is indeed roughly homogeneous. The brightest million galaxies are fairly uniform over the sky (Figure 1.11) and as we look at

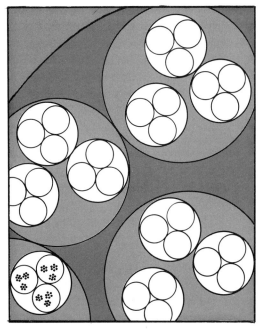

Fig.1.12. Charlier's Universe: an infinite hierarchy of clusters. Such a universe would never look isotropic, however deep our surveys went.

still fainter galaxies, probing to greater distances, clustering becomes less evident and the sky appears smoother. We are not in the kind of Universe discussed by Charlier (Figure 1.12), containing clusters of clusters of clusters *ad infinitum*. Unless we are anti-Copernican and assign ourselves a privileged central position, this apparent isotropy implies that the Universe is roughly isotropic about any galaxy – that the Universe is homogeneous, and all parts evolved in the same way and had the same history.

'The fox knows many things but the hedgehog knows one big thing.' And cosmologists are, in this sense, perhaps, the most hedgehog-like of all scientists. Their subject boasts few firm facts, but each has great ramifications. The first key fact emerged in 1929, when Hubble enunciated his famous law, that galaxies recede from us at speeds proportional to their distance. We seem to inhabit an homogeneous Universe where the distances between any widely separated pair of galaxies stretches as some uniform function of time. This does not imply that we are in a special 'plague spot'. Hypothetical observers on any other galaxy would see a similar isotropic expansion away from them.

Hubble's work suggested that galaxies would have been crowded together in the past, and that there was some kind of 'beginning' ten or twenty billion years ago. But he had no direct evidence for cosmic evolution. Indeed, the steady state theory, proposed in 1948, envisaged continuous creation of new matter and new galaxies, so that despite the expansion the overall cosmic scene never changed.

Origins in a hot 'big bang'

We would not expect to discern any cosmic evolutionary trend unless we can probe at least several billion years back in time. This entails studying objects billions of light years away, which may be invisibly faint even with the largest telescopes. It was Martin Ryle and his colleagues, in Cambridge in the late 1950s, who found the first evidence that our entire Universe was evolving. His telescope could pick up radio waves from some active galaxies (the ones that we now think harbour massive black holes), even when these were too far away to be observed optically. He could not determine a redshift or distance from radio measurements alone, but assumed that, statistically at least, the ones appearing faint were more distant than those appearing intense. He counted the numbers with various apparent intensities, and found that there were too many apparently faint ones (in other words, those at large distances) compared to brighter and closer ones. This was discomforting to the 'steady statesmen', but compatible with an evolving Universe if galaxies were more prone to violent outbursts in the remote past, when they were young.

But the clinching evidence for a so-called 'big bang' came in 1965, when Penzias and Wilson at the Bell Telephone Laboratories detected the cosmic microwave background radiation. This discovery was accidental: their prime motive was a practical one – communication with artificial satellites – and they did not immediately realize what they had found. But the excess background noise in their instruments, which they could neither eliminate nor account for,

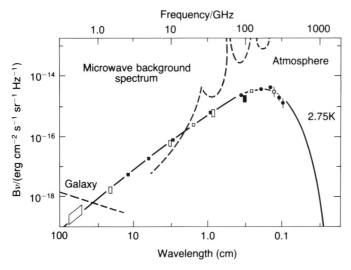

Fig.1.13. The spectrum of the microwave background radiation is best fit by a black body with a temperature of 2.67° above absolute gas. Measurements at frequencies near the peak of the spectrum must be made from above the atmosphere (or indirectly via studies of the excitation of interstellar molecules); at lower frequencies the radiation can be detected at ground level. Penzias and Wilson made their original measurement (marked on the diagram) at about 4 GHz. The temperature is uniform over the sky with a precision of 1 part in 10^4 – this is the best evidence that the overall cosmic expansion is isotropic, and that the Universe is highly homogeneous on the largest scales we can observe (billions of light years).

meant that even intergalactic space is not completely cold: it is about three degrees above absolute zero. This may not sound much, but it implies that there are about four hundred million quanta of radiation (photons) per cubic metre. Indeed, there are 10^8 photons for each atom in the Universe (Figure 1.13).

There is no way of accounting for this radiation, and its spectrum and isotropy, except on the hypothesis that it is a relic of a phase when the entire Universe was hot, dense and opaque. It seems that everything in the Universe once constituted an exceedingly compressed and hot gas, hotter than the centres of stars. The intense radiation in this compressed fireball, though cooled and diluted by the subsequent expansion (the wavelengths being stretched and redshifted), would still be around. The Universe would have become transparent after about a million years, when the temperature fell to 3000 degrees and hydrogen recombined. The photons that Penzias and Wilson detected have travelled uninterruptedly since that time – in other words for about 99.99 per cent of cosmic history (Figure 1.14).

Is there any corroboration of the primordial 'fireball'? According to this concept, when the Universe was only a few minutes old, it would have been at a billion degrees. Nuclear reactions would have occurred as it cooled through this temperature range. These can be calculated in detail. The material

[15]

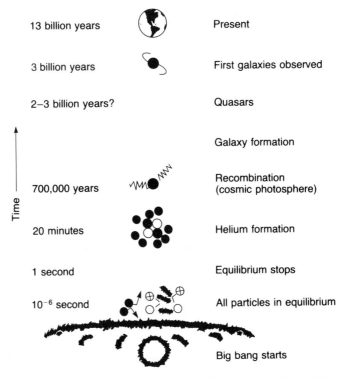

Time		
13 billion years		Present
3 billion years		First galaxies observed
2–3 billion years?		Quasars
		Galaxy formation
700,000 years		Recombination (cosmic photosphere)
20 minutes		Helium formation
1 second		Equilibrium stops
10^{-6} second		All particles in equilibrium
		Big bang starts

Fig.1.14. The history of the Universe according to the so-called 'hot big bang' theory.

emerging from the fireball would be about 75 per cent hydrogen and 25 per cent helium. This is specially gratifying, because the theory of synthesis of elements in stars and supernovae, which works so well for carbon, iron, etc., was always hard pressed to explain why there was so much helium, and why the helium was so uniform in its abundance. Attributing helium formation to the big bang thus solved a long-standing problem in nucleogenesis, and bolstered cosmologists' confidence in extrapolating right back to the first few seconds of the Universe's history, and assuming that the laws of microphysics were the same as now. (Without this firm link with local physics, cosmology risks degenerating into *ad hoc* explanations on the level of 'just so' stories.)

More detailed calculations, combined with better observations of background radiation and of element abundances, have strengthened the consensus that the hot big bang model is basically valid. It is not yet firm dogma. Conceivably, our satisfaction will prove as illusory and transitory as that of a Ptolemaic astronomer who successfully fits a new epicycle – cosmologists are sometimes chided for being 'often in error but never in doubt'. But the hot big bang model certainly seems more plausible than any equally specific alternative – I would personally make the stronger claim that it has more than an even chance of survival.

Consequently, most of us adopt a cosmogonic framework like this. Stars and

[16]

galaxies all emerged from a universal 'thermal soup', at a temperature of ten billion degrees, expanding on a timescale of one second. It was initially almost smooth and featureless, but not quite: there were fluctuations from place to place in the density or expansion rate. Embryonic galaxies – slightly overdense regions whose expansion rate lagged behind the mean value – evolved into disjoint clouds whose internal expansion eventually halted. These proto-galactic clouds collapsed to form galaxies when the Universe was perhaps 10 per cent of its present age; subsequently, the galaxies grouped into gravitation-ally bound clusters. This latter process can be well simulated by N-body dynamical computer calculations. My colleague, Dr Aarseth, has made a movie which illustrates cluster formation. This movie shows the last 90 per cent of cosmic history, speeded up by a factor of about 10^{16}: if the galaxies start off distributed almost uniformly, but with random fluctuations, then regions containing a slightly above-average concentration of galaxies condense into gravitationally bound groups and clusters with a gratifying resemblance to the real groupings seen in our actual sky.

Gravity is crucial here, as it is for the internal evolution within each galaxy. To paraphrase Darwin, it is through the fixed law of gravity that, from amorphous beginnings, structures as varied and wonderful as galaxies, stars and nebulae have been, and are being, evolved. Two features of this peculiar force of gravity are crucial to the whole story.

Gravity drives things *further from equilibrium*. When gravitating systems *lose* energy they get *hotter*. A homely instance of this is the way an artificial satellite speeds up as it spirals downward due to atmospheric drag. Another example is offered by Kelvin's ideas on how the Sun would evolve if its radiative losses were not compensated by the energy from nuclear fusion. A star that loses energy and deflates ends up with a hotter centre than before: to establish a new and more compact equilibrium where pressure can balance a (now stronger) gravitational force, the central temperature must rise. This requires an increase in the star's internal thermal energy, and the star cannot actually radiate more than half the gravitational energy released. This apparent 'antithermo-dynamic' behaviour amplifies density contrasts and creates temperature gradients, a prerequisite for the evolution of any complexity. A one-sentence answer to the question: 'What is happening in the Universe?' might go like this: 'Gravitational binding energy is being released as stars, galaxies and clusters progressively contract, this inexorable trend being delayed by rotation, nuclear energy, and the sheer scale of astronomical systems, which make things happen slowly and stave off gravity's final victory.'

The second key feature of gravity is its *weakness*. The gravitational force within an atom is about 40 powers of ten weaker than the electric force that binds it. But everything has the same sign of 'gravitational charge': it is a long-range force with no cancellation. So gravity holds sway on sufficiently large scales, scales which are vast *because* gravity is so weak. If gravitation were somewhat stronger – 30 and not 40 powers of ten weaker than microphysical forces, for instance – a small-scale speeded-up Universe could exist, in which gravitationally bound fusion reactions had 10^{-15} of the Sun's mass, and lived for less than a year. This might not allow enough time for complex systems to evolve. There would be fewer powers of ten between astrophysical timescales and the basic microscopic timescales for physical or chemical reactions. Complex structures, moreover, could not become very large without themselves being crushed by gravity. Our Universe is vast and diffuse, and evolves so slowly, *because* gravity is so weak. Its extravagant scale, billions of light years, is necessary to provide enough time for the cooling of elements inside stars, and for interesting complexity to evolve around even one star in just one galaxy.

One stumbling-block in understanding galaxies, incidentally, is the rather embarrassing fact that 90 per cent of their mass is unaccounted for. When we study motions in the outer parts of galaxies, and the relative motions of galaxies gravitationally bound in groups or clusters, we infer that the galaxies are feeling the gravitational pull of ten times more stuff than we see (Figure 1.15). There is no reason to be amazed by this – no reason why everything in the Universe should shine conspicuously. What are the candidates for this 'dark

Fig 1.15. A disc galaxy viewed edge on (NGC 4565). Studies of the rotation speeds of gas in the outlying parts of such discs suggest that this gas is 'feeling' the gravitational pull of more material than the stars and gas we actually see. Luminous galaxies are apparently embedded in a more massive and extensive cloud of 'dark' matter. Corroboration of the existence of 'dark' matter surrounding galaxies comes from analyses of the motions of gravitationally bound groups and clusters of galaxies: the total masses inferred from the *relative* motions of galaxies are about ten times higher than those inferred from the internal dynamics of their luminous cores.

matter'? It could be in very faint stars too small to have ignited their nuclear fuel, or, alternatively, in the remnants of massive stars which were bright in the early phases of galactic history but have now all died out. A third idea, much discussed in recent years, is that the primordial fireball might have had extra ingredients apart from the ordinary atoms and radiation that we observe, and that elementary particles of some novel type could collectively exert large-scale gravitational forces.

There are various observational ways of deciding between such varied options. But it would be of special interest to particle physicists if astronomers were to learn more about neutrinos – ghostly and elusive particles which hardly interact at all with ordinary matter – or discover some fundamentally new particles (for instance, the photinos whose existence some theorists predict). We would then have to view the galaxies, the stars, and ourselves in a downgraded perspective. Over four centuries ago, Copernicus dethroned the Earth from a central position. Early this century, Shapley and Hubble demoted us from any privileged location in space. But now even 'particle chauvinism' may have to be abandoned. The protons, nuclei and electrons, which we and the entire astronomical world are made up of, could be a kind of afterthought in a cosmos where neutrinos or photinos control the overall dynamics.

Dark matter – what it is, and how much there is of it – is relevant to cosmogony, especially to the details of galaxy formation. But it is even more crucial for the long-term future – for eschatology. Will our Universe expand forever, and the galaxies fade and disperse into an ultimate heat death? Or will it recollapse, so that our descendants share the fate of an astronaut who falls into a black hole, the firmament falling on their heads to recreate a fireball like that from which it emerged?

Imagine a large sphere or asteroid, which is shattered by an explosion, the debris flying off in all directions. Each fragment feels the gravitational pull of all the others, causing deceleration. If the explosion were sufficiently violent, then the debris would fly apart for ever. But if the fragments were not moving quite so fast, gravity might bind them together strongly enough to bring the expansion to a halt. The material would then fall together again. According to general relativity, this same argument holds for the Universe. In the case of the galaxies, which for the purposes of this argument are envisaged as 'fragments' of the expanding Universe, we know the expansion velocity. What we do not know so well is the amount of gravitating matter tending to brake the expansion. It is easy to calculate how much will be needed to bring it to a halt: it works out at about three atoms per cubic metre. Were the average concentration below this 'critical' density we would expect the Universe to continue expanding for ever, but if the mean density exceeded this value, a big crunch would seem inevitable.

Even if we include the dynamically inferred dark matter in galaxies and clusters, the mean density still falls short of the critical value by a factor of about 5. But there could be some even more elusive material between clusters of galaxies. Until our knowledge of dark matter candidates is less biased and more complete, we will not have a reliable long-term forecast for the Universe.

Martin J. Rees

The ultra-early Universe: initial conditions

But although the alternative long-range futures seem very different, Figure 1.16 highlights a puzzle. The initial conditions that could have led to anything like our present Universe are actually very restrictive, compared to the range of possibilities that might have been set up. We know that our Universe is still expanding after 10^{10} years. Had it recollapsed sooner, there would have been no time for stars to evolve. If it had collapsed after less than a million years, it would have remained opaque, precluding thermodynamic disequilibrium. The expansion rate cannot, however, be too much faster than parabolic, otherwise the expansion kinetic energy would have overwhelmed gravity, and the clouds that developed into galaxies would never have been able to condense out. (This is equivalent to saying that the present density is not orders of magnitude below the critical density.) There is, therefore, a sense in which the dynamics of the early Universe must have been finely tuned. In

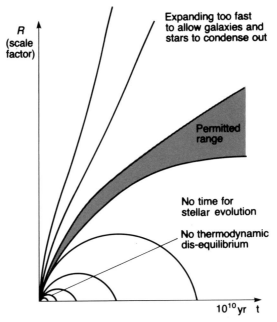

Fig.1.16. This diagram depicts the 'scale factor' for a family of hypothetical isotropic universes which all start off expanding, but are decelerated by the gravitational attraction excited by each part on all other parts. We do not know whether our Universe will go on expanding for ever. We *do*, however, know that our Universe is still expanding, after 10 billion years, but that it is not expanding so rapidly that galaxies could not condense out to form gravitationally bound systems. There is such a sense in which the initial conditions seem to have been 'finely tuned'. Moreover, the initial conditions seem even more special when we note that more general cosmological models (e.g. those which are anisotropic) have other degrees of freedom open to them. Recent ideas in high-energy physics allows us for the first time to approach this question of initial conditions scientifically.

Newtonian terms the fractional difference between the initial potential and kinetic energies of any spherical region must have been very small.

So why was the Universe set up expanding in this special way? And there are other issues that similarly baffle us. Why does the Universe contain fluctuation, while being so homogeneous overall? Why are there 10^8 photons for each particle?

We have pushed the chain of inference right back from the early Solar System to the cosmic fireball at $t = 1$ s. But conceptually, we are in little better shape. We still seem to be appealing to initial conditions: 'Things are as they are because they were as they were.' Our inferences come up against a barrier, just as did the ancient Indian cosmologists who envisaged the Earth supported by four elephants standing on a giant turtle, but did not know what held the turtle up.

Maybe key features of the Universe were imprinted at still earlier stages than $t = 1$ s. The further back one extrapolates, the less confidence one has in the adequacy or applicability of known physics. For instance, the material would exceed nuclear densities for the first microsecond. But if one thinks of time on a logarithmic scale, to ignore these early eras is a severe omission indeed. Theorists differ in how far they are prepared to extrapolate back with a straight face. Some have higher credulity thresholds than others. But those whose intellectual habitat is the 'gee whiz' fringe of particle physics are interested in the possibility that the early Universe may once have been at colossally high temperatures. To motivate this interest, I shall digress for a moment to discuss the basic physical forces.

These forces are just four in number: electromagnetism, the weak force (important for radiative decay and neutrinos), the strong or nuclear force, and gravity. Physicists would like to discover some interrelation between them, to interpret them as different manifestations of a single primeval force. The first modern step towards this unification was the Salam–Weinberg theory, relating the electromagnetic and the weak forces. The basic idea is that at high energies these two forces are the same. They acquire distinctive identities only below some critical energy. Energies in particle physics are measured in giga electron volts, GeV for short, and the critical energy for this unification is about 100 GeV. This energy can just be reached by large particle accelerators, and the Salam–Weinberg theory has been vindicated by experiments at CERN. This development may prove as important in its way as Clark Maxwell's achievement 100 years ago in showing electrical and magnetic effects to be manifestations of a single underlying force (Figure 1.17).

The next goal is to unify the electro-weak force with the strong or nuclear forces – to develop a so-called grand unified theory (GUT) of all the forces governing the microphysical world. But a stumbling-block here is that the critical energy at which the so-called symmetry breaking occurs, the energy which is 100 GeV for the Salam-Weinberg theory, is thought to be 10^{15} GeV for the grand unification. This is a million million times higher than any feasible experiments can reach. It is hard, therefore, to test these theories on Earth. Only tiny effects are predicted in our low-energy world: for instance, protons

Unification scheme for the forces

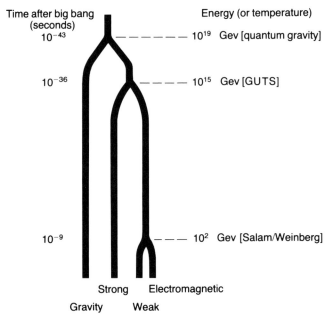

Time after big bang
(seconds)
10^{-43}

Energy (or temperature)

10^{19} Gev [quantum gravity]

10^{-36}

10^{15} Gev [GUTS]

10^{-9}

10^2 Gev [Salam/Weinberg]

Strong   Electromagnetic

Gravity   Weak

Fig.1.17. Schematic diagram showing how, according to some current theories, the four forces of nature are unified at high temperatures which have been attained in the initial instants of the cosmic expansion.

would decay very slowly. But if we are emboldened to extrapolate the big bang theory far enough we find that in the first 10^{-36} s, but only then, the particles would be so energetic that they would all be colliding at 10^{15} GeV. So perhaps the early Universe was the only accelerator where the requisite energy for unifying the forces could ever be reached. However, this accelerator shut down 10 billion years ago, so one can learn nothing from its activities unless the 10^{-36}-s era left some fossils behind, just as most of the helium in the Universe was left behind from the first few minutes. Physicists would seize enthusiastically at even the most trifling vestige surviving from that phase. But it has left very conspicuous traces indeed: it may be that all the atoms in the Universe are essentially a fossil from 10^{-36} s.

If you were setting up a universe in the simplest way, you might make it symmetrical between matter and anti-matter, preparing it with equal numbers of protons and antiprotons. But the particles and antiparticles would then all annihilate as the Universe expanded and cooled. We would end up with blackbody radiation but no matter, no atoms, and no galaxies. However, grand unified theories evade this problem in a way first outlined in a prescient paper by Sakharov, written in 1967. Although these theories predict that proton decay is incredibly slow now, at 10^{-36} s, protons could readily be created or destroyed. As Sakharov first realized, the expansion and symmetry breaking, according to these theories, introduce a slight but calculable favouritism for creation of particles rather than their antiparticles. So that for every 10^8 proton–

antiproton pairs, there is one extra proton. As the Universe cools, antiprotons all annihilate with protons, giving photons. But for every 10^8 photon thereby produced, there is one proton that survives, because it cannot find a mate to annihilate with. The photons, now cooled to very low energies, constitute the three-degree background. There are indeed $\approx 10^8$ of them for every atom. So the entire matter content of the Universe could result from a small fractional bias in favour of matter over antimatter, imposed as the Universe first cooled below 10^{15} GeV.

Grand unified theories are still tentative, but they at least bring a new set of questions – the origin of matter, for instance – within the scope of serious discussion. The realization that protons are not strictly conserved suggests, moreover, that the Universe may possess no conserved quantities other than those, e.g. total electric charge, which are strictly zero. This, combined with the concept of a so-called 'inflationary' phase whereby our Universe could have originated from even a quantum fluctuation, opens the way to envisaging an almost *ex nihilo* origin for our entire Universe.

Frontiers and limits

I began this lecture with phenomena such as ordinary stars where we feel fairly confident that we know the relevant physics. When conditions are more extreme, such as in galactic nuclei, we are less confident, though it is astonishing how far we can go without running up against a contradiction. One theme that has emerged is the interdependence of different phenomena, illustrated picturesquely in Figure 1.18. The everyday world is determined by atomic structure, stars are determined by the physics of atomic nuclei, and the much larger structures (galaxies and clusters) may be gravitationally bound only because they are embedded in clouds of subatomic particles which are relics of a high-energy phase.

But in considering the early big bang, or gravitation collapse inside black holes, we are confronted by conditions so extreme that we know for sure that we do *not* know enough physics. In particular, physics is incomplete and conceptually unsatisfactory in that we lack an adequate theory of *quantum gravity*. The two great foundations of twentieth-century physics are the quantum principle and Einstein's general relativity. The theoretical super-structures erected on these foundations are still disjoint: there is generally no overlap between their respective domains of relevance. Quantum effects are crucial on the microscopic level of the single elementary particle, but gravita-tional forces between individual particles are negligible. Gravitational effects are manifested only on the scale of planets, stars and galaxies, where quantum effects and the uncertainty principle can be ignored. But when the Universe was squeezed to colossal densities and temperatures, gravity could be important on the scale of a single particle, a single thermal quantum. This happens at 10^{-43} s, the Planck time. The effects of quantum gravity would then be dominant, and even the boldest physicists can extrapolate back no further.

Despite the difficulties, some theorists believe it is no longer premature to

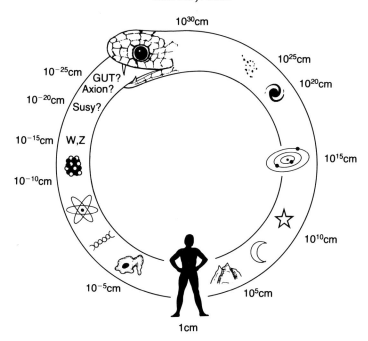

10^{30}cm

10^{25}cm

10^{-25}cm

GUT?
Axion?

10^{20}cm

10^{-20}cm

Susy?

10^{-15}cm

W,Z

10^{-10}cm

10^{15}cm

10^{10}cm

10^{-5}cm

10^{5}cm

1cm

Fig.1.18. New ideas in particle physics suggest further linkages between 'micro' and 'cosmic' scales (i.e. between left and right on this picture) in addition to the well-known ones between atomic and nuclear physics and the terrestrial and stellar scales. The ultimate unification will involve quantum gravity (the Planck scale) and cosmology.

explore what physical laws prevailed at the Planck time, and have already come up with fascinating ideas; there is no consensus, though, about which concepts might really 'fly'. We must certainly jettison cherished commonsense notions of space and time: space–time on this tiny scale may have a chaotic foam-like structure, with no well-defined arrow of time, and there may be no time-like dimension at all. A second idea, generating euphoric optimism at the moment, is that on the tiniest scale space may have extra dimensions. These extra dimensions aren't manifest in the everyday world because they are compactified, rather as a sheet of paper, a two-dimensional surface, might look like a one-dimensional line if rolled up very tightly.

In the light of these speculations about the beginning of time, the demarcation between initial conditions and the laws governing time-development get blurred. Maybe there is only one self-consistent way the Universe could have started off. Our ignorance of the physics governing the first microsecond of cosmic expansion has some analogy with the way ignorance of atomic structure stymied nineteenth-century speculations about the Sun. But there is an important difference. Atoms and nuclei could be probed in the lab, whereas the early Universe is the only place manifesting these ultra-high-energy phenomena. This makes it hard to test unified theories, but at least offers cosmologists a relationship with their physicist colleagues that is symbiotic rather than parasitic.

What is the real chance of clarifying these fundamental questions about physical reality in the next few decades? The prospects are not necessarily hopeless; it isn't presumptuous to try. It is, after all, complexity not sheer size that makes a process hard to comprehend. For example, we already understand the inside of the Sun better than the interior of the Earth. The Earth is more difficult to understand because the temperatures and pressures inside it are less extreme than in the Sun, where it is so hot that everything is broken down into its constituent atoms. For analogous reasons it is harder to understand the tiniest living organism than any large-scale inanimate phenomenon. (Similarly, I have been given an easier task than subsequent speakers in this series.) In the earliest stages of the primordial fireball, matter would surely be reduced, broken down, to its most primitive constituents. So maybe we can realistically hope to understand why the Universe is expanding the way it is, and thereby interrelate the cosmic and microphysical scales in a more profound way. If this goal were achieved, it would supremely exemplify what the physicist Eugene Wigner called the 'unreasonable effectiveness of mathematics in the physical sciences'. It would mean, in a sense, the end of fundamental physics. But this would emphatically not mean the end of challenging science.

A metaphor for what the physicist does is this: suppose you were unfamiliar with the game of chess. Then, just by watching games being played, you could infer what the rules were. The physicist, likewise, finds patterns in the natural world and learns what dynamics and transformations govern its basic elements. In chess, learning how the pieces move is just a trivial preliminary to the absorbing progression from novice to Grand Master. The whole point and interest of that game lies in exploring the complexity implicit in a few deceptively simple rules. Likewise, all that has happened in the Universe over the last ten billion years – the formation of galaxies and the intricate evolution on a planet around at least one star that has led to creatures able to wonder about it all – may be implicit in a few fundamental equations of physics, but exploring all this offers an unending quest and challenge that has barely begun.

Further reading

Shu, F. *The Physical Universe*. University Science Books, 1982.
Hawking, S.W. and Israel, W. (eds.). *300 Years of Gravitation*. Cambridge University Press, 1987.

[2]

Origin of the Solar System

David W. Hughes

The problem of the origin of our planet and the others is one of the fundamental problems of science and is still unsolved. There are two major difficulties. Firstly, we do not know the initial conditions. Coupled with this is the fact that planetary formation took place over 10^9 years ago and many of the chemical and physical processes that occurred in the interim are not time-reversible. So the final state, the Solar System as we see it today, could have been formed in a variety of ways.

The second problem is even more difficult. Only one planetary system can be studied in detail, this being the one we live in. So statistics do not help. Our system and its evolution could be most unusual, a long way from what is the statistical normal. When we examine the characteristics of the Solar System we are hard pressed to distinguish between properties which are probably common to all planetary systems and properties which are peculiar to our own. The situation is akin to that confronting a hypothetical Martian who lands on planet Earth and meets only one person before he is urgently summoned home to Mars by his leaders. He then has to write a scientific paper on the origin of *Homo sapiens*. Say the person he saw was me, a middle-aged university lecturer. Which of my characteristics are common to all Earth beings, which characteristics are important when it comes to the origin of that species? I am male – does he conclude that *all* people are male and that we are sexually hermaphrodite? I am 5 ft 10 in tall with a height-to-width ratio of four – is this important? I have brown hair, is this significant or not? We have the same problem when it comes to our planets. Which facts are important, which are insignificant?

Cosmogonic theories are usually divided into three classes:

   (i) Planets were formed as a direct consequence of the formation of the Sun, the processes being either concurrent or consecutive.
   (ii) Planets were formed after the Sun had become a 'normal' star, out of material dragged out of a passing star or the Sun.
   (iii) Planets were formed after the Sun, out of material captured from interstellar space.

This chapter will discuss the relevant characteristics of our planetary system and newly born stars, will review recently proposed theories and will endeavour to come to some reasonable conclusion.

Introduction

As in all mysteries it is best to concentrate on facts, before diving into the theories. When discussing cosmogony (i.e., the study of the process or processes responsible for the formation of our planetary system), researchers argue about which facts are significant even before they get to the more complex discussions as to the reasonableness of specific theories. So in listing the facts, in what I consider to be an order of importance, I am already encroaching on shaky ground.

Proposed theories are multifarious and are based on such differing concepts as interstellar collisions, accretion from giant molecular clouds, or a simple consequence of single star formation. Reeves has divided the theories into groups according to how the following two questions are answered:

Were the Sun and the planets formed at the same time?
Were the planets formed from interstellar material or from material that has at some time in the past been altered by being incorporated in a star?

I would answer 'yes' to the first and vote for unaltered interstellar material in the second.

At the end we will be left with a 'best buy', but unfortunately one in which we have less than 100 per cent confidence.

When it comes to discussing the origin and evolution of galaxies, stars and planets, the subject which to me is of the greatest interest, the latter, is the one we know least about. The reason is simple. There are 10^{10} galaxies of different sizes, masses and ages. There are 10^{21} stars of different sizes, masses and ages. But there is only one solar system that we know in any detail and that is our own. Recently, we have seen space-probe images of Uranus and its rings and moons. Soon we will be able to sample cometary material and this might answer the second of the two questions above. But when will we know for certain how the planets formed? It's a long time off I'm afraid. What we really need are detailed views of a handful of other planetary systems and who knows when these will be to hand?

The facts

The mean age of the Solar System

Based on the radioactive decay of the element rubidium into strontium (found in Earth, Moon and meteoritic rocks), the mean age of the Solar System is $(4.57 \pm 0.03) \times 10^9$ yr. The age of the Sun is $(5.0 \pm 0.05) \times 10^9$ yr, this rather imprecise value coming from a comparison between present-day solar composition and its assessed original composition. Two things are obvious: within the errors, both figures are the same. So dating does not help us decide whether they were formed at the same time or whether the Sun picked up the planets later on. Also it all happened a very long time ago and the Solar System that we see today could differ considerably from the one that was originally produced. We don't have to go as far as Ovenden who proposed that a planet 90 times heavier than Earth exploded 16 million years ago somewhere between Mars and Jupiter, or Velikovsky who had Venus zooming past Earth at the time of the Noahian deluge. We simply have to look at the structural regularity of the planetary system to realize that gravitational perturbations could have brought this about in 4570 000 000 yr.

Mass and composition

The masses of the planets, their densities and distances from the Sun are given in Table 2.1.

The Sun is 745 times more massive than all the other objects in the Solar System put together. The terrestrial planets are considerably less massive than the major planets but have higher densities. Even when considering uncompressed densities, where allowances are made for the great degree of compression in the interiors of the more massive planets, it can be seen that the planets have differing chemical reactions.

Cosmic material, the material that is used to form stars (and, let us assume, planets too) can be divided into three types, the division being based on melting temperature (see Table 2.2).

From our knowledge of the density of planets and their moments of inertia we can also divide the planetary composition into these three earthy, icy and gaseous types of material. This is done in Table 2.3. The distance of the planet from the Sun, the temperature of the material it condensed out of, and the original mass available, have a considerable influence on the composition of the accreted planet. We can quickly infer that both Jupiter and Saturn are predominantly gaseous.

There are two provisos. Both Jupiter and Saturn emit more radiation than they receive from the Sun and this is thought to be due to the fact that they are still gravitationally contracting, albeit slowly. Measurement of their moments of inertia, using observations of changes in space-probe orbits, indicate that they have a considerable central condensation. Together, these facts point to a rocky core of about 20 earth masses ($M_\oplus$) in both planets. If they were of cosmic composition their present joint masses would have to be about 6000 $M_\oplus$ as opposed to the observed value of 413 $M_\oplus$. So, even though Jupiter and Saturn have accreted a considerable amount of hydrogen and helium from the pre-Solar-System nebula, they certainly did not accrete all that was available. It follows from this that the nebula must have originally been *at least* 15 000 $M_\oplus$, i.e. 10^{32} g in mass (remember that the present solar mass is only 20×10^{32} g).

If condensation takes place at temperatures in the 1000–2000 K range, both the ice and the gaseous material are lost and the rocky metallic planetesimals that do accrete can subsequently be accreted to produce the terrestrial planets and the asteroids.

[28]

Table 2.1. *Planetary data*

	Orbital semi-major axis (AU)	Inclination of orbit to ecliptic (deg)	Eccentricity	Mass (10^{27} g)	Density (g cm^{-3})	Uncompressed density (g cm^{-3})	Rotation period	Inclination of equator to orbital plane
Sun	–	–	–	1 989 000	1.41		25.38^d	(7.25)
Mercury	0.39	7.00	0.206	0.330	5.40	5.4	58.65^d	0
Venus	0.72	3.40	0.007	4.870	4.20	~5.2	−243.01^d	2
Earth	1.00	–	0.017	5.980	5.518	~4.2	23.9345^h	23.44
Mars	1.52	1.85	0.093	0.642	3.95	3.3	24.6229^h	23.98
Asteroids	2.70	~9.50	~0.140	>0.0048	~3.50		8.00^h	
Jupiter	5.20	1.30	0.049	1899.450	1.34		9.84^h	3.08
Saturn	9.54	2.48	0.056	568.640	0.70		10.23^d	2.90
Uranus	19.18	0.77	0.047	86.90	1.58		15.50^h	97.92
Neptune	30.06	1.77	0.009	102.970	2.30		(15.8)^h	28.80
Pluto	39.44	17.17	0.025	0.011	~0.50		6.39^d	(> 50)

Table 2.2. *Cosmic material, the 'stuff' of stars and planets**

	Earthy	Icy	Gaseous
Elements and Compounds	Si, Mg, Fe, S, etc. plus O	C, N, O plus H	H, He
Melting point	~ 2000 K	< 273 K	< 14 K
Available mass	1	2.2	200

Note: (*See Table 2.7.)

Table 2.3. *The estimated composition of the planets*, in percentage form*

	Earthy	Icy	Gaseous
Terrestrial planets	100	< 1	0
Jupiter	6	~13	~81
Saturn	21	~45	~34
Uranus	~28	~62	~10
Neptune	~28	~62	~10
Comets	~31	~69	~ 0

Note: (*Jupiter and Saturn have rocky cores, both of about 20 $M_\oplus$. Uranus has a theoretically estimated rock and ice core of about 13 $M_\oplus$.)

Table 2.4. *The rebuilding of the solar nebula*

	Mass ($M_\oplus$)	Condensation zone (AU)	Enhancement factors to achieve solar composition		
			Weidenschilling	Harris	Whipple
Mercury	0.053	0.22–0.56	500	300	500
Venus	0.0815	0.56–0.86	290	300	500
Earth	1.000	0.86–1.26	320	300	500
Mars	0.107	1.26–2.06	250	300	500
Asteroids (P)	1.3×10^{-3}	2.00–3.30		(300)	(500)
Jupiter	318.0	3.3–7.4	2–40	3	10
Saturn	95.0	7.4–14.4	10–60	5	30
Uranus	14.6	14.4–24.7	50–140	30	75
Neptune	17.2	24.7–35.5	50–115	30	75
Pluto	0.0026	35.5–45.0		(30)	(75)

Uranus, Neptune and the comets seem to be very similar in composition. If you condense those components of the solar mixture that freeze out at temperatures around 100 K, the end result could simply be a collection of comets. Allow a large number of comets to coalesce and you accrete Uranus and Neptune.

This concept of having a solar mix of materials in each of the zones before the planets start to come together enables us to reconstruct the solar nebula. The present composition of each planet is used as a starting-point and it is then

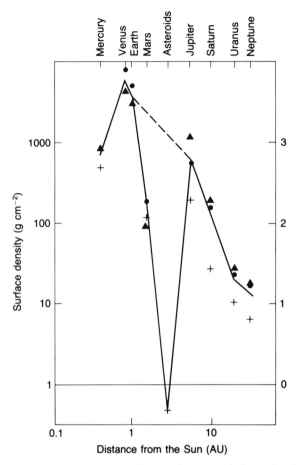

Fig.2.1. The surface density of the pre-planetary nebula as a function of heliocentric distance. The bold line is obtained by joining the arithmetic means of density estimates made by Weidenschilling ▲; Whipple ●; and Harris +. The dashed line is an interpolation between Earth and Jupiter.

augmented to the solar composition by assuming that no iron was lost during the accretion process. This enhanced mass is then spread out over the zone of the nebula that condensed to produce the planet, and a surface density is calculated. The procedure is best illustrated by an example. Earth has an iron mass fraction of about 0.38. Bringing Earth up to solar composition would augment the mass a total of 320 $M_\oplus$. This is then spread over an annular zone of inner radius 0.86 AU and outer radius 1.26 AU, giving a surface density of 3200 g cm^{-3} (an AU is the mean Earth–Sun distance).

A similar calculation can be made for the other planets (see Table 2.4). Figure 2.1 shows a plot of mean surface density of the resulting pre-planetary solar-system nebula as a function of distance from the Sun.

It would be dangerous to draw too sweeping a conclusion from this diagram, but four points seem worth making. First the decrease in density as the

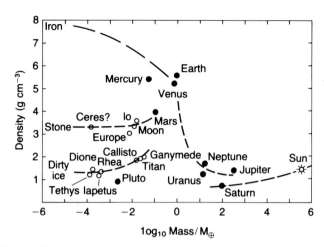

Fig.2.2. The densities of planets and satellites in the Solar System plotted as a function of their mass (given logarithmically in terms of earth masses).

heliocentric distance increases beyond Jupiter's orbit is consistent with the planetary system terminating at Neptune. There is little material left over for a Planet X and it probably doesn't exist. (Note that I am completely disregarding Pluto. This object, with a mass of around a quarter of the mass of the Moon is very unplanetlike and is probably an escaped satellite.) Second, moving to the inner Solar System, it is highly probable that there is no intramercurial planet. Third, it is apparent that something very odd happened in the asteroid belt to cause the 'Cassini division' in the Solar System and some mechanism must be invoked to explain the sweeping away of a considerable percentage of the mass that used to reside in this zone; not just the ice and gaseous components were lost but obviously much of the iron and rock too. The fourth point concerns the mass of the pre-planetary nebula. Today's Solar System has a mass of 0.001 34

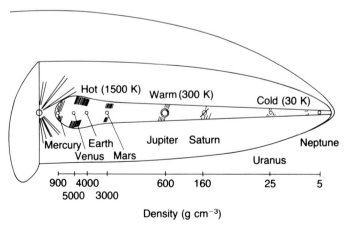

Fig.2.3. The probable temperature and density of the flattened nebula dust ring during the first stages of planetary accretion.

times the solar mass. The nebula it condensed out of could easily be a factor 500 times more massive, a typical suggested mass being 0.5 solar masses.

The planetary and satellite composition variation can be summarized in Figure 2.2. Certain specific density classes can be clearly seen. The low-density 'dirty ice' satellites, Dione, Rhea, Callisto, Ganymede, etc., follow a smooth curve, the density increasing by compression as the mass increases. Stony objects, Ceres, Moon, Io (Mars) behave similarly. The terrestrial planets obviously consist of stone *and* iron, Mercury's density clearly showing that there was a concentration of iron and nickel in the inner part of the solar nebula before condensation. The major planets have densities scattered around the mean solar density *and* that of the dirty ice.

The density variation has been taken by many cosmogonists to be an important clue to the temperature variation in the pre-planetary nebula. This is illustrated diagrammatically in Figure 2.3.

Angular Momentum

It has already been stated that the Solar System has a mass which is only 0.13 per cent of the solar mass. The picture is entirely reversed when it comes to angular momentum. The angular momentum of the Sun is about 1.7×10^{48} c.g.s. units. The large majority of the angular momentum of a planet is orbital and its numerical value is found by multiplying the planet's mass by both its mean orbital velocity and the mean planet–Sun distance. Of the total angular momentum of the whole system, Sun plus planets, 61 per cent resides in Jupiter. Saturn, Uranus and Neptune account for a further 38 per cent and the Sun has only 0.5 per cent. This fact has worried cosmogonists for decades. Why is the Sun spinning so slowly? How quickly would we expect it to be spinning? To answer this we must turn to the interstellar medium. Stars condense out of dense clouds of interstellar matter. Once the condensation becomes controlled by its own gravitation its angular momentum remains constant, thus:

$$\frac{\text{initial period of rotation}}{\text{final period of rotation}} = \left(\frac{\text{initial diameter}}{\text{final diameter}}\right)^2$$

The present-day solar diameter is 1.39×10^{11} cm. Following Hoyle, the cluster in which the Solar System was born was typically 15 light years across and broke up into 1000 stars with initial interstellar distances of about 10^{18} cm. Gravitational independence occurs at, say, one-tenth of this value, so 10^{17} cm is a reasonable estimate for the initial diameter in the above equation. The Sun orbits the galaxy in about 2×10^8 yr. Clouds which are condensing to form stars spin faster than this and a reasonable rotation period is 3×10^7 yr. So the final spin period of the Sun, using these values and the formula given above, is 0.02 d. Its present-day equatorial spin period is 25.3 d, the equatorial velocity being 2 km s^{-1}. The above calculation leads to an equatorial velocity of 2400 km s^{-1}. In fact, the present Sun could not spin at this speed as it would become unstable at 400 km s^{-1}. The fast spinning condensation mentioned above would produce a lenticular primordial Sun with an equatorial disc of radius two-fifths that of Mercury's orbit.

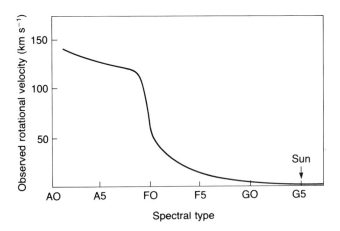

Fig.2.4. The observed rotational velocity of main sequence stars as a function of spectral type (after Bernacca and Perinotta).

If we follow the condensation carefully, it is interesting to note that at a diameter of 1.5×10^{14} cm (the diameter of Jupiter's orbit) the Sun's equatorial spin period is 73 yr. Jupiter's orbital period is 11.9 yr, these two values being reasonably close from an astronomical standpoint. So the planets are orbiting as fast as is to be expected. It is the Sun that has slowed down. And the Sun is not unusual in this respect. Figure 2.4 shows how the equatorial spin velocity of main sequence stars varies as a function of spectral type. The sharp drop between F0 and F5 has been attributed to two causes. The first was the angular momentum transfer from the star to its surrounding planets during the T Tauri phase of stellar evolution. This has now dropped from favour because the planets have no excess angular momentum. The second process that could cause a drop in rotation velocity is the removal of angular momentum by the stellar wind. In the T Tauri stage of the Sun's evolution, this wind was passing Earth at a velocity of about 450 km s^{-1} and was dense, the Sun losing mass at a maximum rate of about 10^{26} g yr^{-1}. This T Tauri phase lasted around 50 million years, during which time the Sun's radius decreased from nearly that of the radius of Mercury's orbit to its present 7×10^5 km. Also, the Sun lost a considerable amount of mass. (It is quite reasonable to conclude that the Sun commenced its T Tauri phase with twice the mass that it now has.) This solar wind is capable of sweeping clean the newborn solar system and blowing away the 'leftover' nebula agreement, which was made up mainly of volatile elements such as hydrogen and helium, the elements that did not condense into the inner planets and only partially condensed into the outer ones.

Planetary rotation data is given in Table 2.1 and this can be both summarized and compared to asteroid data by plotting spin angular momentum as a function of mass, as shown in Figure 2.5. The spin angular momentum, H, of a body is simply the product of its angular spin frequency, ω, and its moment of inertia, I. (The moment of inertia is given by $I = Mr^2\alpha$ where M is the mass, r the radius and α is a function of the density distribution and can be taken as 0.4

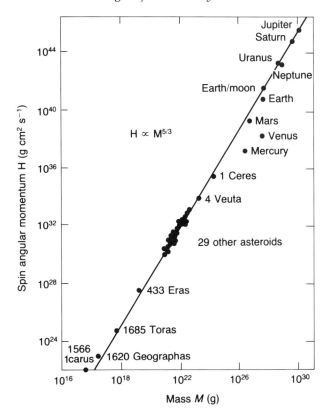

Fig.2.5. A logarithmic plot of the spin angular momenta of planets (with the exception of Pluto) and 35 astroids with well-determined sizes, shapes and rotation rates as a function of their masses (after Fish, Hartmann and Larson, and Burns).

for a solid equidense sphere.) The straight line in Figure 2.5 has the form $H = 1.47 \times 10^{-5} M^{5/3}$ where H and M are both in c.g.s. units. The rotational period, T, is found by substituting $T = 2\pi/\omega$ into the above formulae. This gives $T = 4.3 \times 10^{5} \alpha r^2 M^{-2/3}$. If the body is assumed to have a constant density, ρ, throughout (i.e. $M = 1.33 \pi r^3 \rho$) and an α of 0.4 then $T = 6.59 \times 10^{4} \rho^{-2/3}$, which, when the assumption is made that all Solar System bodies have the same density and no central mass concentration, confirms Alfvén's findings that all planets and asteroids have approximately the same period of rotation. Using a density of 3 g cm^{-3} this period comes to 8.8 h.

One possible reason for the constancy of spin period is apparent if it is assumed that all planets and asteroids have been formed by accretion. This process becomes less and less efficient as these bodies get nearer and nearer to their rotational stability limit. This limit occurs when the equatorial rotational surface velocity of the body is equal to its escape velocity (i.e., when $T \approx 3.3\rho^{-1/2}$ h, a rotational period of ~ 2 h for a body of density 3 g cm^{-3}). A conclusion might be that bodies cannot accrete if they spin faster than about one-third the rotational stability limit.

[35]

It can be seen from Figure 2.5 that Mercury and Venus have considerably lower angular momenta than is predicted by the linear relationship. This is probably due to long-term tidal interactions between these planets and the Sun. The high-spin deceleration suffered by Venus (changing its period from ~8 h to 243 d) indicates that tidal dissipation inside the planet was considerable. This would have been possible only if large regions in the interior of Venus were near their melting point, thus enabling creep elastoviscosity and direct viscous interactions to play their part in dissipating energy. It is not beyond the bounds of possibility that Mercury was once in orbit around Venus.

Earth and Moon are also way below the line in Figure 2.5, but if the angular momentum of the spinning Earth is combined with the orbital momentum of the Moon, the Earth–Moon system moves up onto the linear relation. This indicates that the angular momentum of the total system has changed little during the history of the Solar System. All that has happened is that the Moon has moved further away and the Earth now spins more slowly.

Two other characteristics are worth mentioning. The large majority of Solar-System objects are spinning directly, in the same sense as their orbital motion. If the ring of planetesimals is dense, as observations indicate, the gravitational interactions between the particles would lead to the disc revolving as a solid body, the orbital velocity increasing as one moves away from the Sun. The resulting planet would spin directly.

The orientation of the spin axes is not perpendicular to the ecliptic plane (i.e., the plane of the Solar System). These deviations are probably due to statistical fluctuations in the final accumulation of any planet. In other words Uranus's unusual 98° obliquity could be produced by the angular momenta brought in by the last few accreting planetesimals (see Figure 2.6).

The spin axis of the Sun is 7.25° away from the normal to the plane of the Solar System. Maybe a close stellar passage caused this deviation. Note, in passing, that the plane of the Solar System is inclined at 60° to the plane of the galaxy.

Size and regularity

The furthest major planet from the Sun, Neptune, has an orbital semi-major axis of 30 astronomical units (AU), a figure which should be compared to the distance of the present-day closest star to the Sun which is 270 000 AU. So the known Solar System is very small.

Is there a 'Planet X' beyond Neptune? Figure 2.1 indicates that the density of the nebula is decreasing quickly in the outer regions so there is very little material present to form an outer planet. Also, the time taken to accrete a planet from a collection of planetesimals varies as a function of distance from the Sun. Hoyle calculated that it would have taken about 2×10^6 yr to form the terrestrial planets and about 300×10^6 yr to form Uranus and Neptune. The hypothetical Planet X would have required even longer and we have to remember that the Solar System is only about 4570×10^6 yr old.

Remember, too, that the distance between the Sun and its nearest stellar

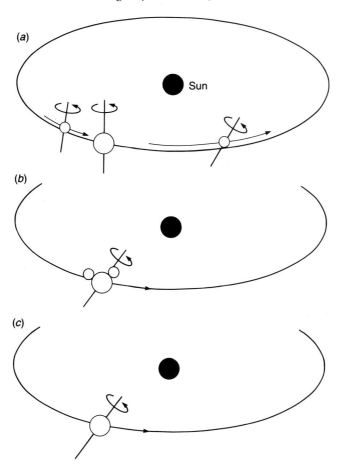

Fig.2.6. The accretion of the last few planetesimals dictates the direction of the final spin axis. This axis will precess due to both the accretion process and the torque exerted on the not-exactly-spherical final planet.

neighbour is always changing. The number of stars, N, that have passed within a distance, d, of the Sun during the lifetime of the Solar System, T_0, is given by

$$N = \pi n V T_0 d^2,$$

where n is the spatial density of stars (at present about 0.097 pc^{-3}) and V is the mean relative velocity between the Sun and its neighbours (about 20 km s^{-1}). Substituting into the above equation gives

$$N = 6.7 \times 10^{-7} d^2 \ (d \text{ in AU}).$$

So if the mean stellar spatial density remains at its present value the closest a star is likely to have been to the Sun during the age of the Solar System is about 1200 AU, 40 times further away than Neptune. When the Sun was in its natal open cluster, things could have been more crowded.

The regularities of the Solar System have been a matter of considerable interest for a long time. Historically, we saw Johannes Kepler trying to fit the five regular solids, the cube, tetrahedron, dodecahedron, icosihedron and octohedron into diminishing exscribed spheres, these spheres having the orbits of Saturn, Jupiter, Mars, Earth, Venus and Mercury as equatorial great circles. The Titius–Bode law, in which the radius of the orbit of the nth planet was given by

$$r_n = 0.4 + 0.3 \times 2^n \text{ AU}$$

($n = 0$ for Venus, 7 for Neptune, Mercury having $r_n = 0.4$), was also stressed. Table 2.1 shows that, with the exception of Mercury and Pluto the two planets at the extremities, the orbits are coplanar, direct and circular. None of these facts are of great significance. They simply underline the age of the system and the cumulative effect of mutual gravitational perturbation. Also, in an accretion process, where the majority of particles are moving directly, the particles having retrograde and/or eccentric orbits will have a much shorter time interval between collisions and will preferentially be eradicated.

The evolution of the Sun

The origin of the Solar System must be synchronized with specific stages during the evolution of the Sun. Dense, cool clouds of gas molecules and dust condense to form a cluster of protostars. Small density perturbations are unavoidable in these clouds and under specific conditions the action of gravity will make the small perturbations grow, forcing the cloud to break up into a number of condensations which eventually become stars. The conditions are given by the Jeans Criterion. Consider a cloud of radius R and density ρ. If the total energy of the cloud is negative it will contract under its own gravitation. The cloud has two types of energy, a gravitational energy equal to $-GM^2/R$, where G is Newton's constant of gravity and M is the cloud mass, and a thermal energy of $1.33\, A\rho T\pi R^3\mu^{-1}$ where A is 8.3×10^7 erg mole^{-1}deg^{-1}, T is the cloud's absolute temperature and μ is the molecular weight of the gas. A cloud with radius greater than R_1 will collapse if

$$R_1 \approx \frac{0.2}{T}\frac{M}{M_\odot}\text{pc}$$

(one parsec is 3.086×10^{13} km or 3.26 light years).

So an ordinary interstellar cloud with $M \sim M_\odot$ (the solar mass) and $R \approx 1$ pc will not collapse. A dense gas-dust cloud with $M = 10^3$–$10^4 M_\odot$, $T = 10$–50 K and $R \approx 10$ pc will condense and will form an open cluster of 10^4–10^5 stars. Luckily, contraction does not automatically lead to a large rise in temperature because collisional excitation of the rotation levels of hydrogen molecules followed by the emission of 28 μ infrared radiation easily removes most of the thermal energy. The cloud condenses until its density has risen to the point at which it becomes opaque to its own infrared radiation. Then the temperature of the central region of the star grows rapidly. The subsequent evolution has been

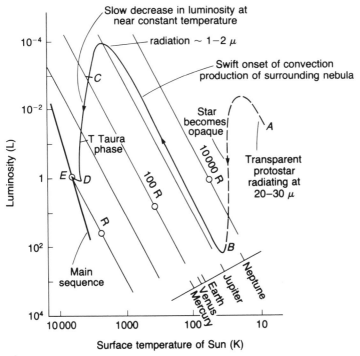

Fig.2.7. The evolutionary track across the Hertzsprung-Russell diagram of a star which finally ends on the main sequence with a mass of 1 $M_\odot$.

modelled theoretically by Chushiro Hayashi who found that the gravitational energy released by the contracting protostar is transported to the surface by convection. Near the surface this convective energy is converted to radiation at an equilibrium temperature of about 2500 K. When convection is first established the luminosity of the star flares to a maximum value, L, given by

$$L \approx 10^4 \left(\frac{M}{M_\odot} \right)^2 L_\odot,$$

this maximum lasting for a few years. After this the star's radius will continue to decrease, the surface temperature remaining around 2500–3500 K. This results in a steady decrease in luminosity. During this period the central temperature steadily rises, passing quickly through a value of several hundred thousand degrees, a temperature at which a nuclear reaction involving lithium, beryllium and boron occurs. Eventually it reaches a point at which the proton–proton reaction takes place. At this time the generated thermonuclear energy produces a sufficiently high temperature for the gas and radiation pressure to balance the gravitational force. The star stabilizes, its position on the Hertzsprung–Russell diagram marking a point on the main sequence. The evolution track is shown in Figure 2.7. Two stages of speedy contraction occur and in both the gas and dust is undergoing a free fall. The first, from A to B on Figure 2.7, is accompanied by a flare of infrared radiation in the 20–30μ range. The second contraction coincides with the star becoming opaque. This causes a

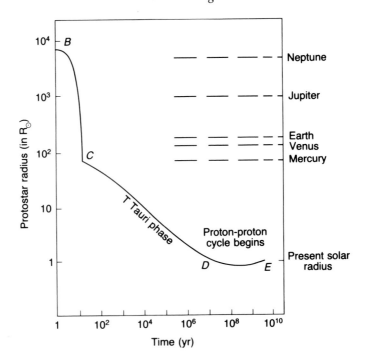

Fig.2.8. The radius of a protostar as a function of time (after G.H. Herbig).

massive rise in the central temperature, leading to a break-up of all the molecules into ions and electrons and to the onset of convective heat transfer. This is accompanied by a flare of near infrared radiation and a decrease in radius from a value which is a few times $10^3 R_\odot$ to a figure of about $10^2 R_\odot$ (B to C on Figures 2.7 and 2.8). This takes place extremely quickly in astronomical terms (i.e., a few tens of years), and it can be seen from Figure 2.8 that during this period the radius of the Sun drops from a value equivalent to the radius of Neptune's orbit to that of Mercury's. This fact is of significance when it comes to theorizing about the origin of the planets because during this short time the Sun flares in luminosity to a value of about 10^4 times its present value, the associated radiation pressure helping to detach the outer layers of the Sun which are then left behind as the core collapses. Theoretical studies indicate that the collapse of the protostellar cloud is extremely non-homologous, so that the cloud quickly develops a high-density core which accretes material from the outer regions. These outer regions form the nebula from which the planets subsequently condense. The role of angular momentum and magnetic fields during the collapse is crucial in determining whether the cloud eventually forms a multiple star system, a star with planets or a single star.

In the penultimate stage (C–D in Figures 2.7 and 2.8) the Sun is a T Tauri type star and during this period its radius decreases by a factor of ~ 100 and its luminosity by $\sim 10^3$. This phase lasts about 10^8 yr and in this period the luminosity varies irregularly at optical wavelengths. The underlying late F to

[40]

middle M spectral class star is surrounded by an active chromosphere ($T \approx 2 \times 10^4$ K, density 10^9–10^{12} atoms cm^{-3}) which produces many emission features. The star is losing mass, the largest rate of mass outflow occurring when it is young and still confined within a dark nebula cloud. Mass outflow decreases as the star approaches the main sequence and it is possible that this mass outflow is coupled to the photospheric layers by strong magnetic fields, thus providing the mechanism whereby the star loses the large majority of its angular momentum and slows down to a rotation period of a few tens of days. The mass outflow can typically be 10^{-5}–10^{-6} M$_\odot$ yr^{-1} (the Sun at present is losing 10^{-14} M$_\odot$ yr^{-1}). The velocity is several hundred km s^{-1} and this wind is probably sufficient to halt further accretion from the protostellar cloud.

S. Strom states that 'during the first stages of dissipation, light from the newly formed star "leaks" out through small breaks in the placental dark cloud and illuminates patches of more distant dark cloud material. These illuminated patches are identified with the Herbig–Haro objects'. A bright pre-main sequence, the T Tauri star, in the constellation of Lupus, the Wolf (the star known as RU Lupi) has been found to be attended by a swarm of dark orbiting protoplanets. G. Gahm and his colleagues observed the star photometrically and spectroscopically using three telescopes simultaneously and found that the spectral lines were completely unaltered during the brightness fluctuations. This indicates that flaring of the underlying star was not responsible for the variation. Intrinsic mechanisms modify the spectrum so the responsible agent must be external to the star. The astronomers concluded that dust concentrations in orbit around the star periodically blocked out the light trying to reach Earth. These absorbing clouds were found to have dimensions of around 10^5–10^6 km, masses of the order of 10^{21}g (i.e., similar to asteroids) and orbits typical of inner planets.

The conclusion is that RU Lupi has in orbit around it a collection of condensing protoplanets which will eventually shrink and accrete to form planets.

The prevalence of planetary systems

We know for certain that there is a planetary system around the Sun, but what about other nearby stars? Detection is the problem. A single star is seen from Earth to move in a simple curve across the sky. Stars with planetary systems should reveal themselves by having irregular 'wobbly' celestial motions. What is seen from Earth is the movement of the centre of gravity of the star plus planets and this can be interpreted in terms of the mass and mean orbital radius of the components. Van de Kamp at Sproul Observatory has studied nearby stars, searching very carefully for irregularities in their movement. Of the stars closer than 4 pc, four (other than the Sun) have planetary systems. These are listed in Table 2.5 and shown in Figure 2.9.

So of the 32 stars in this restricted volume of space, 4 out of the 16 single stars have planetary systems (one for certain and three suspected) and 1 out of the 8 binary systems has a component with suspected planetary companions.

Table 2.5. *Four stars within 4 pc of the Sun are suspected of having planetary systems*

Name of star	RA(1950) h m	Dec (1950) (°)	Distance away (light years)	M_v	Spectral type	Approximate stellar mass (solar masses)	Approximate mass of planet(s) (Jovian masses)
Barnard's Star	17 55.2	+ 4 33	5.9	13.2	M5	0.11	1.1
							0.8
Lalande 21185	11 00.6	+36 18	8.1	10.4	M2	0.23	10.0
ε Eridani	3 30.6	− 9 38	10.7	6.1	K2	0.71	6.0
61 Cygni	21 04.7	+38 30	11.2	A 7.5	K5	0.49	8.0
				B 8.4	K7	0.39	–
Sun	–	–	–	4.83	G2	1.0	1.4

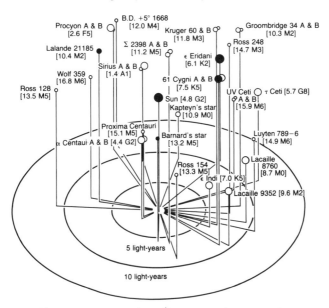

Fig.2.9. There are 31 stars within 13 lt yr (4 pc) of the Sun, of which 16 are present as 8 binaries. The absolute visual magnitude and spectral class is given in parentheses after each star name. The filled circles represent the stars which are thought to have planetary systems (adapted from Colin A. Roman, *Deep Space*, Macmillan, 1982, p. 53).

Bearing in mind the difficulty of detection, it would be wrong to draw sweeping conclusions from these data. But still it is obvious that the ratio 4:16 points to planets being common companions for stars of spectral classes G,K and M and it is reasonable to assume that planet formation is a common but not essential adjunct to star formation (see Figure 2.21).

Meteorites

The Earth is continually being peppered by interplanetary objects, some of which survive the passage through the retarding atmosphere and end up as meteorites which can be carefully analysed in the laboratory. The large majority of meteorites originate from the asteroid zone and are collision fragments of a few large asteroids. Meteorites can be divided crudely into irons and stones. The stones (carbonaceous chondrites, ordinary chondrites and achondrites) contain minerals such as olivine and troilite which condensed out of the nebula at high temperatures, irregular matrix material which is a low-temperature condensate from the vapour state, and chondrules, millimetre or so sized spherical rocks, which seem to have condensed from the liquid state. These condensates can be understood in terms of the large range of pressures and densities that existed in the inner regions of the Solar System nebula.

Iron meteorites are composed of a nickel–iron alloy and have a structure indicative of being cooled from around 800 K to 500 K at a very slow rate, a few degrees per million years. For this to occur the iron must have been enclosed in a much larger parent body and was probably part of the core of a large asteroid. Some of the rocky meteorites could represent the segregated crust of such an object. So the parent body was heated during formation, subsequently cooled slowly and then suffered a collision-induced fragmentation which dispersed the meteorites into a range of orbits, some of which intersected the Earth's orbit.

Special mention must be made of the Allende meteorite which fell in Mexico in February 1969 and was found to contain anomalous amounts of the isotopes Mg^{26} and Al^{26}, both of which have half-lives for radioactive decay of around 7×10^5 yr. The only known source of these is remnants of supernova explosions. So a supernova explosion must have occurred within a hundred parsecs of the Solar System around the time of its origin.

This is completely unsurprising. The Sun, as stated on p.33 started life as a member of an open cluster of around 10^4 stars. The handful of massive stars in this cluster would move through to the fourth stage of their evolution in a few times 10^7 years. Supernovae explosions would help them lose mass as they collapsed towards the white dwarf region of the main sequence and many of the heavy elements that had been produced in the star by nuclear synthesis would be spewed out into the cluster. The Sun takes 10^8 years to get to the main sequence and would find *its* nebula cloud liberally contaminated by these heavy elements. As we will see, the shock wave produced by the supernova explosion might even have helped trigger the condensation of the planets.

Planetary formation

It will have become obvious from the preceding pages that the favoured theory for the origin of the Solar System has the origin of planets associated very closely with the origin of the Sun. The abundance of radioactive isotopes such as rubidium 57, thorium 232 and uranium 238 dates the solidification of Earth, Moon and meteorites at around $(4.57 \pm 0.03) \times 10^9$ yr before the present. Measurement of plutonium 244 (half-life 8×10^7 yr) and iodine 129 (half-life 1.6×10^7 yr) indicate that the protosolar material had become isolated from the interstellar medium at most 10^8 yr before planetary formation. This can be taken as a large hint that the Sun and planets have a quasi-contemporary formation. This dual event can be placed in a galactic context by referring to Figure 2.10. A cloud of interstellar material, the protostellar nebula, i.e., future Sun and planets, orbits and galactic nucleus, in the plane of the galactic disc, every 200 million years, the orbital radius being about 10 000 pc. As the cloud passes from an interarm region into an arm, it is seriously decelerated, causing considerable contraction. At each contraction there is a finite probability of star formation being triggered. The galactic arms are regions where stellar formation and nuclear synthesis is occurring. Radioactive plutonium 244 and iodine 129 added to the cloud at point 1 in Figure 2.10 will have decayed by the time the cloud reaches point 2. If we assume that the protosolar nebula started to collapse at point 2, this nebula could have picked up radioactive material during its first 2×10^7 yr. Notice that the massive shortlived stars completely

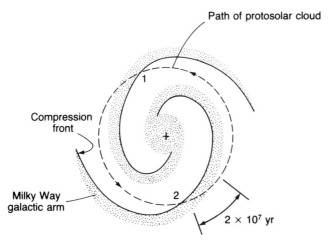

Path of protosolar cloud

Compression
front

Milky Way
galactic arm

2 × 10⁷ yr

Fig.2.10. The path of the protosolar cloud around the galactic nucleus is shown as a dashed circle which has a radius of about 10000 pc. Star formation can occur when the cloud enters the compression front. The dotted galactic arm is a region containing many new stars, the more massive of which quickly become supernovae and seed the region with radioactive heavy elements and shock waves. If the protosolar nebula started to condense at point 2, the Sun would be nearing the end of its T Tauri phase by the time it had reached point 1.

evolve from protostars through their supernova phase before they have left their natal galactic arm.

Three problems are encountered in going from cloud to star: the final remnant is too hot, too highly magnetized and is spinning too rapidly. The first problem is easily overcome because the molecules efficiently radiate away the thermal energy. The galactic arms have a magnetic field of about 3 microgauss. Contraction progressively increases the cloud's field strength, which opposes the shrinking. During contraction the rotation increases. Luckily, the cloud's magnetic field is still connected to that of the galactic arm. Cloud rotation accelerated by the contraction effectively drags the field lines with the plasma and the field lines resist stretching. Thus magnetic and rotational energy is transferred from the rotating mass to the galactic arm and the condensing cloud becomes lenticular, as shown in Figure 2.11. Some of these discs would have a central mass concentration, others would have a more uniform density distribution. A.G.W. Cameron suggested that the first type condensed to form one star plus a planetary system, whereas the second type led to binary stars.

The mechanism for stopping the condensation of the lenticular disc was triggered by the transition from an isothermal to an adiabatic process. The disc was approximately isothermal during the early stages when it was opaque. It was also reasonably isothermal during the second stage of its evolution, a time when the released potential energy was used to evaporate compounds such as H_2O, NH_3 and CH_4 from the dust grains, and to dissociate and subsequently ionize these compounds. As soon as there was little left to absorb this energy the temperature and, concomitantly, the pressure increased sharply, leading

[45]

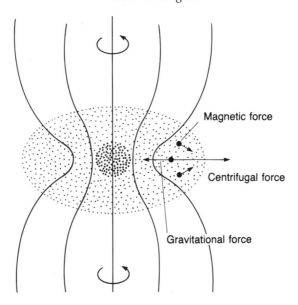

Fig.2.11. The condensing protosolar cloud is represented by the dotted regions, the black lines being magnetic field lines. The interplay between magnetic, centrifugal and gravitational force produces the lenticular shape. The deflection of the field lines and the magnetic gradient so created leads to a decrease in the rotational and magnetic energy of the condensing cloud.

to a termination of the collapse. At this stage the nebula would have been about 30 AU in radius. The central regions would have had a temperature of a few × 10^4 K and the cloud was still at a temperature of above a few × 10^3 K beyond the present orbit of Jupiter. Over a period of time (see the region *A–B–C* in Figure 2.7) the disc separated into two discrete components, a protosun (radius equal to that of Mercury's orbit) surrounded by a less dense tenuous nebula of dust and gas. In the central region of the nebula the high temperature (10^4 K) had vaporized all the dust. But as the Sun retreated, the region cooled and the metallic elements started to condense out. Moving away from the infant Sun the temperature dropped even lower and progressively refractory and earthy materials, and then volatile substances such as the ices of water, ammonia and methane, condensed. In the outer regions many of the dust grains even retained their interstellar form. This disc is spinning and the particles have orbits of varying sizes, eccentricities and inclinations. Their velocities can, however, be resolved into two components, a general movement around the Sun and a much smaller chaotic motion between individual particles. Particles collide frequently, these collisions being non-elastic, resulting in energy loss. The component of the velocity perpendicular to the mean plane decreases more than the component in the plane, the result being that the disc flattens, as shown in Figure 2.12.

As the dust particles are more massive than the gas molecules and atoms, the dust has far smaller chaotic velocities. Dust thus retreats to the equatorial plane

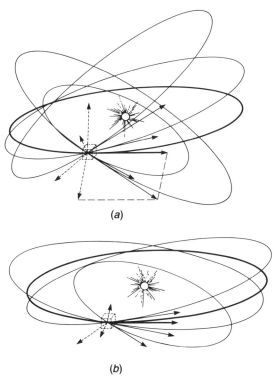

(a)

(b)

Fig.2.12. The flattening of the orbiting cloud of particles as their chaotic movement decreases. Thin arrows show the velocity of separate particles, thick arrows the velocity of the general circular movement, and the broken arrow, the component of the velocities of the particles (from Boris Levin, *The Origin of the Earth and the Planets*, Foreign Languages Publishing House, Moscow, 1956).

of the flattened cloud forming a flat evolving disc as shown in Figure 2.13.

We are now confronted with a rather difficult problem. How do we convert this dense, slowly rotating disc, with its myriad of small (less than a millimetre) particles and its envelope of gas, into the planetary system that we see today? There are two main possibilities. Firstly the dust disc could remain unbroken during the period when its surrounding gas is being blown away by a solar wind mechanism (this wind rising to a gale during the T Tauri phase of the Sun). Or the disc could fragment and undergo self-gravitational collapse, the nebula subdividing into numerous balloons of collapsing gas and dust – each orbiting the infant sun and having differing compositions and gas/dust ratios as a function of their heliocentric distance. This second scenario could continue with the dust–rock–ice falling to the centre of each balloon and condensing to form a protoplanet, with a typical dimension to that of our Moon. These subsequently collide suffering either accretion or fragmentation until, eventually, the Solar System is populated with the present selection of planets.

To decide between the two possibilities we have to return to the Jeans Criterion and we must also consider the added problem of tidal disruption.

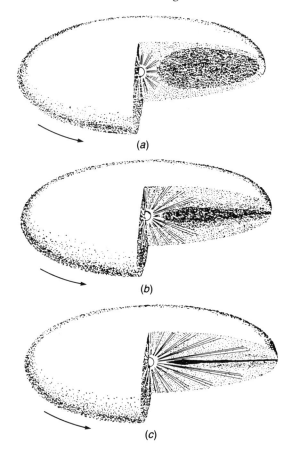

Fig.2.13. In the spinning preplanetary solar nebula the dust retreats to the equatorial plane due to the energy losses that occur in the collision process illustrated in Figure 2.12 (from Boris Levin, *The Origin of the Earth and the Planets*, Foreign Languages Publishing House, Moscow, 1956).

The Jeans Criterion governs condensation. Consider a fragment of the gas–dust disc mass M_f, radius R_f, temperature T_f and density ρ_f. This fragment can only condense if its gravitational energy is greater in modulus than its thermal energy, that is,

$$\frac{3}{5}\frac{GM_f^2}{R_f} > \frac{3}{2}RTM_f$$

where G is Newton's universal gravitational constant and R is the universal gas constant. If we replace the radius by the density we get

$$M_f > \left(\frac{5}{2}\frac{RT}{G}\right)^{3/2}\left(\frac{4}{3}\pi\rho_f\right)^{-1/2},$$

i.e.

$$M_f > 8.5 \times 10^{22}T^{3/2}\rho^{-1/2}.$$

[48]

The total disc is about 30 AU in radius. Perpendicular to the plane, only pressure balances the gravitational force so the thickness of the disc is about $\frac{1}{40}$ of its radius. A mass of around 10^{32} g will give it an average density of about 1.4×10^{-11} g cm^{-3}.

So
$$M_f > 2.3 \times 10^{28} T^{3/2} \text{ g.}$$

The disc temperature T ranges from 2000 K down to, say, 10 K, so the onset of condensation is somewhat touch and go and requires very cool regions with densities considerably above the average to trigger it.

Tidal disruption tends to occur because the Sun attracts the sunward side of the condensate more than the anti-sun side and this tends to pull the condensate apart. Opposing this is the gravitational attraction between the different regions of the condensate. Condensation can only occur if

$$\frac{\beta M_f}{R_f^3} > \frac{M_\odot}{r^3},$$

where β is a dimensionless constant which has a value somewhere between 1 and 3. The quantity r is known as the Roche limit, the distance from the Sun beyond which condensation can occur. It is given by

$$r > \left(\frac{0.12 M_\odot}{\rho}\right)^{1/3}.$$

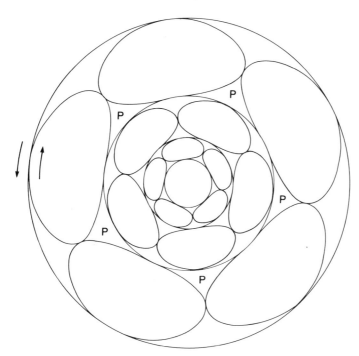

Fig.2.14. A plan view of the equatorial disc of the nebula showing the vortex pattern proposed by von Weiszäcker. Dust and gas concentrations occur at P.

Using a density of 1.4×10^{-11} g cm^{-11} gives $r > 17$ AU, which again poses a problem because this is nearly equal to the size of the orbit of Uranus.

I will mention three solutions that have been put forward in an attempt to solve this problem. All essentially rely on increasing the density in the equations above. In Figure 2.13(c) the thin gas and dust equatorial disc is spinning in such a way that collisional friction is tending to make the angular velocity decrease as one moves further away from the Sun. These differences in orbital period are a constant source of turbulence, a turbulence which has a tendency to break the disc into a series of whirlpool-like eddies. In a flat disc the turbulence will be rather gentle and the vortices will tend to have their axes perpendicular to the equatorial plane. Figure 2.14 shows the vortex pattern proposed by von Weizsäcker. Most turbulence tends to increase viscosity and this usually leads to a breakdown of any specific pattern and a subsequent increase in disorder. We must not forget, however, that the Sun continues to pump energy into the system and so von Weizsäcker's vortices can be compared to the large-scale circulation patterns in the Earth's atmosphere. The gas and dust becomes concentrated in the inter ring regions (marked P in Figure 2.14) and this increase in density triggers condensation and the formation of planetesimals. Von Weizsäcker suggested that there were five vortices simply because the separations of the patterns gave a tolerable

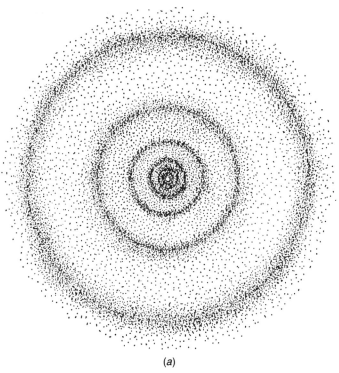

(a)

Fig.2.15. (a) The gas concentrations shown in Figure 2.14 merge and the equatorial disc of the nebula resolves into a set of concentric rings.

approximation to the Titus–Bode law of planetary distances. The end product is a series of rings which can subsequently accrete to form planets (see Figure 2.15).

A second scenario has been put forward by Williams. He regarded the nebula equatorial disc as being similar to the rings of Saturn. The larger objects in the disc set up a series of resonances and this, coupled with a tendency for

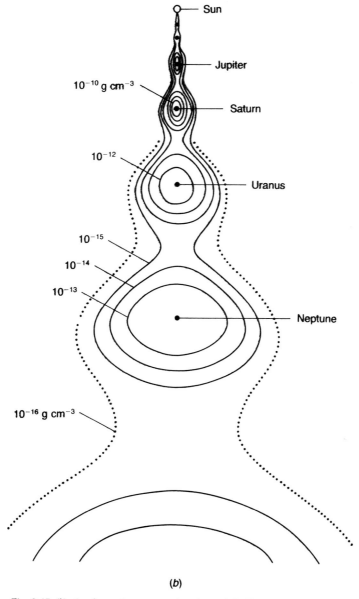

(b)

Fig. 2.15 (b). A schematic cross section through half of the disc shown in (a). The lines represent density contours.

[51]

the orbits to circularize, resulted in an uncomfortably long period being required for the accumulation of a planet. Something is needed to produce a drastic increase in the density and this is where the supernova that injected the Allende meteorite's Al^{26} into the nebula comes in handy. Many link the onset of star formation with the passage of a supernova-generated shock wave. Maybe a subsequent supernova shock wave could trigger planetary condensation in the flattened disc. For a typical supernova the ratio, u, of the shock speed to the speed of sound in the nebula is about 5000. This produces a compression ratio, x (the density in front of the shock divided by the original density), of

$$x \sim 0.4u^{6/5}, \quad \text{i.e., } x \sim 10^4.$$

This large increase in density drops the Jeans mass by a factor of 100 and changes the Roche limit from 17 AU to 0.80 AU. The result is the formation of a collection of gas and dust protoplanets. The grains settle to form a central core and the increase in temperature resulting from this is sufficient to expel some of the low molecular weight peripheral gases (i.e., the hydrogen and helium).

The third approach to the problem concentrates on the very small. Micron-sized dust grains are unaffected by the Roche limit due to their small size, high density and internal strength. Grain–grain collisions are commonplace. Gravitation is far too weak to assist accretion. There is, however, no reason to suppose that the dust behaves like billiard balls. It is most likely to be sticky. Ice at temperatures below its melting point certainly is. And laboratory experiments simulating molecular cloud chemistry often produce a series of organic tar-like materials which could act as ideal glues.

In this model everything starts with dust. The dust grows by accretion. Differing-sized dust particles are then produced which themselves stick together when they collide, forming even larger objects. A typical end product is shown in Figure 2.16. Note that we deliberately leave out the size of this object, it could be anything from a centimetre to tens of kilometres.

Whichever process one chooses, the end result is the break-up of the equatorial disc (Figure 2.13(c)) and the subsequent production of a multitude of planetesimals – some formed by accretion, others by direct condensation. These planetesimals have a range of sizes from tens of metres up to tens of kilometres and also a range of compositions, these echoing the temperature of the nebula, which decreases drastically with heliocentric distance.

Then follows a period of collisions, collisions that result both in growth and fragmentation. We know that growth won in the end; the planets are there to prove it! But the accumulation was an extremely wasteful process. From a nebula of cosmic composition and mass somewhere in the 0.1–1.0 solar mass region the Solar System residue is a mere 0.0013 solar masses.

Collisions in this disc persist as long as there is appreciable non-conformity in the velocities of the components, but these non-conformities will soon die down, due to the impact conditions shown in Figure 2.12(b). Eventually, the planetesimals will be moving in nearly circular orbits and only inter-planetesimal gravitational forces can cause accretion. Now, the gravitational

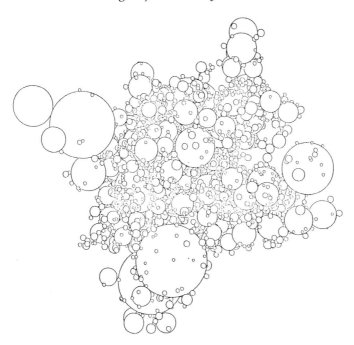

Fig.2.16. A 'fractal' model of an aggregate in the Solar System. The picture has been considerably simplified. If you take it to be a model of a 1-km diameter planetesimal each individual sphere is itself an aggregate which essentially looks like the whole.

forces between objects which are a few metres in diameter are very weak but this must be compared with the difference in the solar gravitational force acting on two neighbouring orbiting objects. This is also very small. In a few thousand years the metre-sized objects will have accreted to form planetesimals with sizes in the 1–100 km range. The process progressively slows down. Aggregation depends on the number of collisions taking place, and this is a function of the orbital parameters and the number of bodies present.

The mean orbital parameters will fluctuate. Initially, when the concentric rings (see Figure 2.15(a)) have been established the particle orbits will be regular, as shown in Figure 2.17(a). Accretion will decrease the number of objects in the ring. Gravitational perturbations will generate radial motions and motions perpendicular to the equatorial plane (giving inclinations up to a few degrees). The volume of space through which the bodies move increases and probability of collision drops. Hoyle has considered the timescale of this process and his results are summarized in Table 2.6.

As the accretion continues, the remaining few bodies in the ring region find themselves in orbits which are less and less likely to lead to a collision. The time taken for each stage increases progressively, eventually approaching infinity. If this was not the case the end product of the Uranus–Neptune ring would be one coagulated planet and not the two which are observed to be there at

[53]

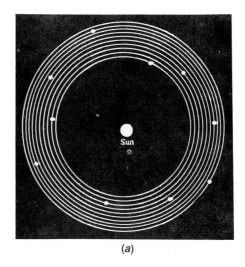

(a)

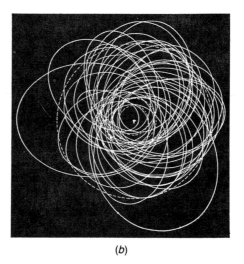

(b)

Fig.2.17. During the accumulation of Uranus and Neptune there is stage (*a*) when around 50 bodies of mass 4×10^{27} g (and many smaller ones) are orbiting in the zone. The circulation looks regular but is in fact somewhat erratic, rather akin to a stock car race. Gravitational perturbation randomizes the orbits, see (*b*) during which time accretion slows down. Many of the smaller bodies are ejected at this stage (after Fred Hoyle, *The Cosmology of the Solar System*, University of Cardiff Press, 1978, Figures 9.1 and 9.2).

present. These two planets are clearly moving in near circular orbits so the stage illustrated in Figure 2.17(*b*) will be followed eventually by a much more regular very-few-object scenario. The transition from the ring shown in Figure 2.17(*a*) to the ring shown in Figure 2.17(*b*) is accompanied by considerable mass

Table 2.6. *Aggregation*

Mass of largest planetesimals (g)	No. of bodies	Timescale of aggregation (yr)	Total mass of ring region (g)
(A) The aggregation of Uranus and Neptune			
10^6	10^{25}		10^{31}
10^{14}	10^{16}	10^1–10^2	10^{30}
5×10^{21}	4×10^7	10^3	2×10^{29}
4×10^{27}	50	10^6–10^7	2×10^{29}
10^{29}	2	3×10^8	2×10^{29}
(B) The aggregation of the inner planets			
10^6	10^{22}		
6×10^{17}	2×10^{10}	$> 10^2$	
6×10^{25}	200	$> 10^4$–10^5	
6.6×10^{27} (Earth–Mars)		$> 2 \times 10^6$	
5×10^{27} (Venus–Mercury)	2		

loss, and it can be seen from Table 2.6 that as the mass of the largest planetesimal grows from 10^6 g to 10^{14} g to 5×21^{21} g the total mass in the ring volume is progressively decreasing. In the Uranus and Neptune area the planetesimals are made of a mixture of dust and ice, the ice mainly being H_2O but containing small percentages of CO_2, NH_3 and CH_4. The swarm of objects with masses in the range 10^{14}–10^{22} g is essentially indistinguishable from a great stream of cometary nuclei. Many collect to form Uranus and Neptune, but these growing planets also perturb many of the remainder. Some are ejected from the Solar System and wander off in the plane of the galaxy. Others find themselves losing energy and being transferred into typical short-period cometary orbits in which they speedily decay, breaking up into dust streams and feeding the zodiacal dust cloud. A small percentage are given orbits which take them into regions between the Sun and its near-neighbour stars. Here the dirty snowballs are effectively stored in the deep freeze of space. Occasional perturbations from passing stars push them into orbits which bring them close to the Sun again and at that time they decay, produce comas and tails, and can be seen as comets. The present-day collection of comets has a total mass of around that of the Earth. This is the exponential tail of what was a much larger cloud.

The accreting planetesimals originally have the temperature of the ambient nebula. On colliding with and accreting to the growing protoplanet they bring to it their potential energy. Imagine that an object like that shown in Figure 2.16 grows by accretion until it has a mass M. Its potential energy, V is given by

$$\frac{V}{M} = 0.8 G \pi \rho R^2,$$

where ρ is its density and R its radius. Assume, as a limiting case, that all this energy is converted into heat. The temperature rise, ΔT, so generated is given by

$$\Delta T = \frac{0.8 G \pi \rho R^2}{C_p} ,$$

where C_p is the specific heat of the accreting substance ($\sim$ 800 J kg^{-1}C^{-1} for rock).

For planet Earth, the resulting ΔT is 45 000 K, for an asteroid of radius 100 km the value is 11 K. Remember that this is a maximum value and that radiation will reduce it considerably, but still the value obtained for Earth is sufficient to make the rock viscous if not molten and this will enable the gravitational field to pull the planet into a spherical shape. The low temperature found in the case of the asteroid indicates that no melting or rock flow takes place. Small asteroids remain irregular objects.

Many of the processes mentioned are summarized in Figure 2.18. The Sun evolves through its T Tauri stage to its present position on the main sequence. The gaseous nebula, some of which condenses on the colder accreting planetesimals, is in the main simply blown away from the Solar System by the intense stellar gales of the T Tauri stage and the flaring in the early main sequence life of the Sun. The composition of the original nebula, both gas and dust is given in Table 2.7. Here we have taken the cosmic abundance of the elements in the early Sun and its vicinity from Allen. These elements can be crudely divided into the three building blocks of the planetary system: (i) rock and iron, (ii) ice and (iii) gas. The mass ratios are 1:2.2:200. So no planet has a cosmic composition. This can be illustrated by considering Jupiter and Saturn. Grossman and colleagues found that both have rocky–iron cores of mass around 20 Earth masses. If they had cosmic composition their total masses would be 203.2 times larger, i.e., around 4000 Earth masses. In reality, they have masses of 318 M$_\oplus$ and 95 M$_\oplus$, respectively, so gas (hydrogen and helium)

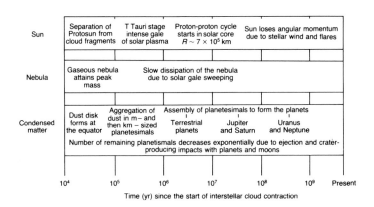

Fig.2.18. A summary of the timescale of major events in the early history of the Sun and planets.

[56]

Table 2.7. *Cosmic abundance of the elements*

Atomic no.	Element	No. atoms per 10^6	gms per tonne	Atomic no.	Element	No. atoms per 10^6	gms per tonne
1	H	920 466	735 000	12	Mg	24	474
2	He	78 340	249 000	14	Si	30	685
8	O	608	7710	16	S	15	377
10	(Ne)	76	1220	28	Ni	2	86
7	N	84	950	13	Al	2	44
6	C	304	2920	11	Na	2	30
26	Fe	37	1645	20	Ca	2	58
				18	(Ar)	6	184

If these are crudely divided into metal, rock, ice and gas (assuming 50% of Fe is present as metal and three atoms of O combine with every two of Fe, Mg, Si, Al, etc. in rock) the masses per tonne are

metal	rock	ice (H_2O, CO_2, NH_3, CH_4)	gas (H, He)
910 g	4000 g	11 000 g	984 000 g

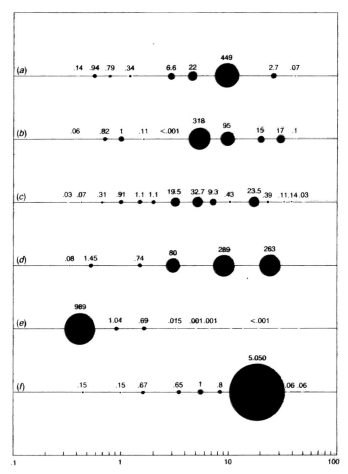

Fig.2.19. Alternative Planetary Systems (see Carl Sagan The Solar System in *Scientific American* September 1975, **233** (3), 31). Numbers above the hypothetical planets denote mass in multiples of the Earth's mass. The horizontal scale indicates the semimajor axes of the planets' elliptical orbits. The system labelled (b) is our Solar System.

has been lost in profusion. Also, the loss mechanism seems to be more effective the further one is from the Sun. The T Tauri gale and subsequent solar wind and flare have been suggested as the cause. Any mechanism resulting in the heating the nebula will also suffice.

The velocity V_e required to escape from the Sun's gravitational field is given by

$$V_e = \left(\frac{2M_\odot G}{r}\right)^{1/2}$$

where $M_\odot$ is the mass of the Sun and r is the distance from its centre. The mean random velocity, c, of a collection of gas molecules is given by

$$c = \left(\frac{3RT}{\mu}\right)^{1/2},$$

where R is the universal gas constant and μ is the gram molecular weight. The gas will escape in around 50 000 yr if $c > V_e/4$ and for this to be the case for molecular hydrogen, at the orbital distance from the Sun of Jupiter, Saturn and Uranus, requires the temperature of the nebula to be 4425, 2300 and 1000 K respectively.

Nor should we be left with the impression that our Solar System is the only one that could have formed out of the condensing equatorial ring. Theoretical modelling of the accretion process, obviously a hit-and-miss business, has led to the six systems shown in Figure 2.19. The variability is startling.

Other ideas

Modern astronomical observations of the arms of our galaxy indicate that there is an intimate relationship between the cloud-like concentrations of interstellar material and the places where stars are formed. The majority of present-day theories concerning the origin of the Solar System have the Sun and the planets condensing out of the same cloud of gas and dust and the origin of both occurring at the same time. A large condensing cloud breaks up into many fragments due simply to it having too much angular momentum. This is shown schematically in Figure 2.20. If we take the small fragments shown in Figure 2.20(*d*), the end point of their evolution also depends rather drastically on their angular momentum. Three possibilities are shown in Figure 2.21. Due to gravitational instabilities it is probable that single stars are more likely to have planetary systems than binary systems. Even so, it is reasonable to conclude that around one in four of the stars in the Milky Way are blessed with planetary systems.

Let us briefly move back through astronomical history. Over the centuries many theories for the origin of the Solar System have been suggested and it is very difficult to eliminate any one of them completely. Firstly, there are far too few criteria which are discriminatory. Secondly, all models are blessed with elements of flexibility and as such can easily adapt to accommodate new observations. Thirdly, the subject is extremely complex and encompasses a vast range of physical sciences ranging from geochemistry to plasma physics, orbital dynamics to statistics. Most of these sciences are encountered in conditions remote from the laboratory.

Theories divide into cogenic and non-cogenic. Laplace propounded one of the former in his 1796 book on popular astronomy, *The System of the World*. His theory remained popular for nigh on 150 years. It is illustrated in Figure 2.22. As the protostellar cloud contracted its rotation increased and the cloud became lens-shaped, gravitational force balancing the centrifugal force. The cloud was only able to contract further after it had allowed an equatorial ring of material to split away. As contraction continued it again became equatorially unstable and peeled off another ring. This went on until the remainder of the remainder of the material fell to form the Sun, which stabilized at its present size. The rings subsequently agglomerated to form planets. The death knell of the Laplace theory was angular momentum. In order that the primordial

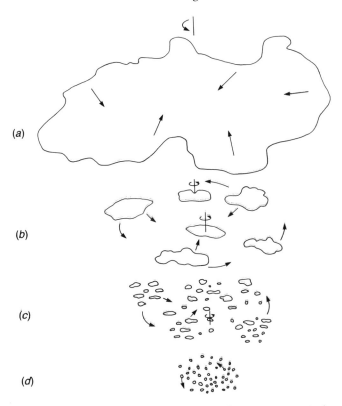

Fig.2.20. As an interstellar gas and dust cloud collapses it progressively fragments and subfragments transferring most of its spin angular momentum into the orbital angular momentum of the resulting stellar-mass parts. When it has reached (*d*) it is similar to an open cluster.

nebula could flatten and lose rings it needed to be spinning very quickly. The resultant central Sun should be spinning in hours and not days. Conversely, imagine the Sun extended out to fill the dimension of the present planetary system: rotation would be so slow that ring detachment would be out of the question.

Sir James H. Jeans propounded a tidal theory in which the substance from which the planets were made was torn out of the Sun by the gravitational attraction of a massive star speeding past. The torn-off material was drawn away from the Sun and compelled to orbit the Sun in the plane containing the passing star's orbit. The angular momentum problem is solved by adjusting the nearest encounter distance. The cigar-shaped filament of attracted material (see Figure 2.23) subsequently broke up into 'drops' which condensed to form planets.

One rather dubious attribute of this theory was that the suggested phenomenon was extremely rare. One would not be confronted with a superabundance of planetary systems with the attendant development of life, civilization

Fig.2.21. The fragments of the original gas–dust cloud condense to form differing protostellar nebula dependent on their angular momentum. Low angular momentum (A) produces a large protostar and a low-density nebula, the end result probably being a single star with no planets. A medium amount of angular momentum (B) results in a much more massive nebula and the end product is a star plus a planetary system. A fragment with high angular momentum (C) produces a binary or multiple star system (after J.A. Wood *The Solar System*. Prentice-Hall, 1979, Figure 7.1).

Fig.2.22. Laplace's hypothesis of planetary formation.

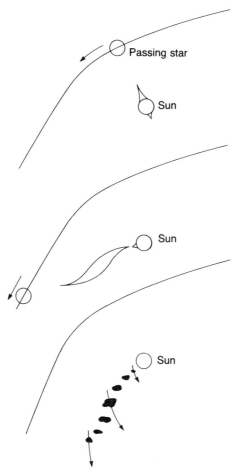

Fig.2.23. Jeans' tidal theory of planetary formation. The passing star drawing a filament of material from the Sun, this filament subsequently condensing to form planets.

and religion. Scientifically, it ran into even more serious problems. It is impossible to give material large amounts of angular momentum without also giving it large amounts of energy. Provide the present Solar-System angular momentum and the material escapes from the Sun. If the substance is given less than the escape velocity, it finds itself in eccentric orbits which eventually collide with the Sun. Also, if sufficient material is pulled away it has to come from deep solar regions where the temperature is high (10^6 K). The 'drops' subsequently expand much more quickly than they can cool and all the gases disperse into space.

A much more plausible tidal theory has recently been proposed by Woolfson. Here the early Sun, immersed in its natal open cluster, encounters a still-condensing protostar. The protostar, mass ⅟₇ the solar mass, radius 3×10^9 km (20 AU), density 4×10^{-12} g cm^{-3}, temperature 30 K passes within 45 AU of

David W. Hughes

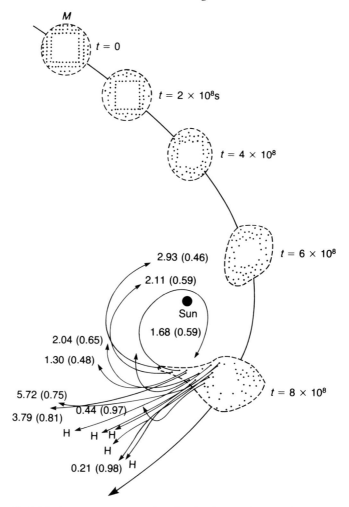

Fig.2.24. A protostar passes close by the Sun. Orbits marked H are hyperbolic. The remainder is captured, the numbers marked on the orbits being perihelion distance (units of 10^{14} cm) and in brackets the orbital eccentricity (after Woolfson 1964).

the Sun. A computer model of this distorted protostar is shown in Figure 2.24. Under specific circumstances, the tides exerted on the ejector by the Sun and the passing protostar virtually cancel out, leaving the cold droplets free to condense into planets.

The origin of the Moon has always been a matter of considerable interest and a common proposal in the early part of the twentieth century relied on another form of condensation-induced break-up of a spinning fluid object. A typical scenario is shown in Figure 2.25. Unfortunately the mass ratio between Earth and Moon is about 81 to 1 and Jeans found that break-up in a rotating fluid commonly led to mass ratios around 10 to 1. It might be no coincidence that the mass ratio of Earth:Mars is 9:1 and of Venus:Mercury 15:1.

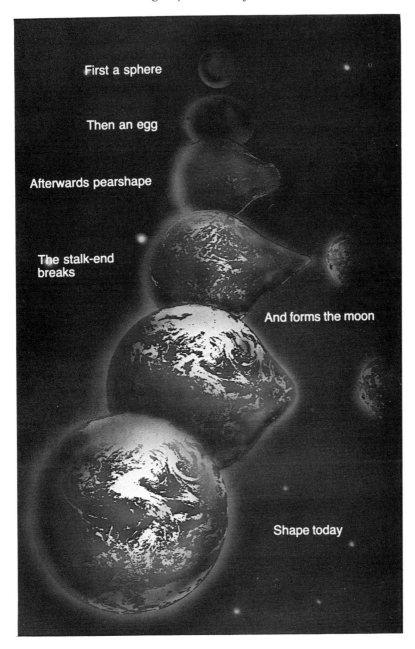

First a sphere

Then an egg

Afterwards pearshape

The stalk-end breaks

And forms the moon

Shape today

Fig.2.25. A protoplanet spins faster as it contracts. The fluid object changes shape, necks and breaks up into two major fragments which tend to have a mass ratio of 10:1. Smaller fragments may be left behind at the break point. Maybe Earth:Mars formed this way (Mass ratio 94:1). The Moon (1/81 the Earth's mass could have been one of the small fragments) (from *Splendour of the Heavens*, eds. T.E.R. Phillips and W.H. Steavenson, Hutchinson, 1923, p. 3).

[65]

Let us end this review with a brief description of an accretion theory. Interstellar space contains large amounts of gas and dust and the Sun, on its 200-million-year orbit around the galactic nucleus, passes from time to time through these gas–dust clouds. The idea is simple, the Sun is formed at one specific time and picks up a nebula sometime later. This nebula then condenses to form the planets. Lyttleton, Schmidt and Pendred and Williams are the main proponents of accretion theories. If the Sun, mass $M_\odot$ moves at a velocity V through a gas cloud of density ρ_c it will accrete mass at the rate dM/dt given by

$$\frac{dM}{dt} = \frac{4\alpha\pi G^2 M_\odot{}^2 \rho_c}{(V^2 + W^2)^{3/2}}$$

where W is the mean speed of the gas molecules in the cloud, relative to the star ($W = 2(2kT/\pi m_g)^{1/2}$ where k is Boltzmann's constant, T is the absolute

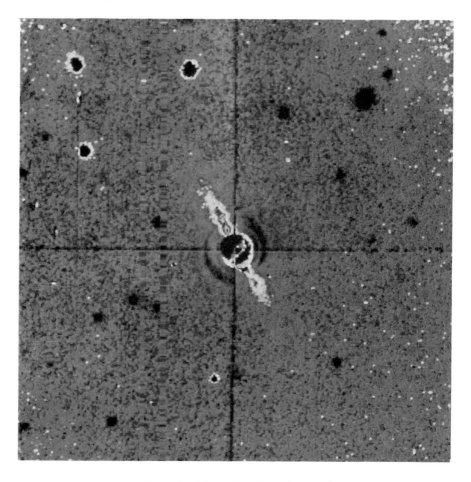

Fig.2.26. The IRAS image of a protoplanetary stellar nebula surrounding the star β Pictoris. The nebula extends for about 400 AU on each side of the star (courtesy University of Arizona and JPL).

temperature and m_g is the mass of the gas molecule (or atom)). A typical hydrogen gas cloud has a temperature of a few hundred Kelvin and typical values of V and W are about 1 km s^{-1}. So dM/dt = 2.5 × 10^{35}ρ_c g s^{-1} and to accrete the planetary mass (10^{31} g) from a cloud of ρ_c10^{-21} g cm^{-3} would take about 10^6 yr, a not unreasonable time period. Unfortunately under simple conditions all the accreted material falls straight into the Sun. This is not the case, however, if the cloud has some angular momentum or if the accretion is asymmetric due to the presence of another nearby star.

Conclusion

Cosmogony, the study of the origin of the Solar System, is not an easy subject. Essentially, if we are allowed to choose special conditions independent of their rarity, there are many ways in which we can end up with a set of four terrestrial planets and four gaseous planets orbiting our star Sun. With a sample of one, statistics are of little help. So as a last resort let us lean on those two old standbys of natural philosophy, the simplicity rule and Ockham's Razor. Nature is pleased with simplicity and entities should not be multiplied unnecessarily. Let us do away with colliding stars, penetrated gas clouds and Laplacian ring formation and let us be satisfied with planetary formation as a simple adjunct to star formation. If we have too little angular momentum in the condensing gas–dust cloud we end up with a single star, too much, and we produce a binary, but if, like baby bear's porridge, it is just right, we get one star plus a series of planets. This will happen in about one in five cases. The Milky Way galaxy contains the Sun and around 100 000 000 000 other stars. Divide this number by five and you have the number of solar systems. This is not the place to go into the complexities of counting the number of planets that are suitable for life or the fraction of those on which life actually evolves and becomes intelligent and manipulative. But still the conclusion is that planets are common.

The Infra Red Astronomy Satellite has a 0.6-m telescope on board that was used to image the star Beta Pictoris (see Figure 2.26). Surrounding this star is a flattened gas–dust nebula similar to the nebula illustrated in Figure 2.13(*b*). All we have to do is to observe for the next 100 million years or so and the Beta Pictoris planetary system will slowly accrete before our eyes.

Further reading

Allen, C.W. *Astrophysical Quantitites*. 3rd edition, Athlone Press, 1973.

Dermott, S.F. (ed.) The Origin of the Solar System. John Wiley & Sons, 1978.

Evrett, E.H. (ed.) *Frontiers of Astrophysics*. Harvard University Press, 1976.

Hoyle, F. *The Cosmogony of the Solar System*. University College Cardiff Press, 1978.

[67]

David W. Hughes

Kuiper, G.P. (ed.) *The Atmospheres of the Earth and Planets*. University of Chicago Press, 1949.
Wood, J.A. *The Solar System*. Prentice-Hall, 1979.

[3]

Origins of complexity

Ilya Prigogine

The historian A. Koyré has aptly described the western scientific revolution of the seventeenth century as a transition 'from the closed world to the infinite universe'. Still, classical science concentrated its efforts mainly on stable, periodic motions. We are now beginning to understand the message of the second law of thermodynamics: we are living in a world of unstable dynamical systems, and it is this very instability which leads to stochastic and irreversible processes. Some of the main subjects of contemporary science, be it biology or cosmology, illustrate in a striking way the constructive role of irreversible processes. We must abandon the myth of 'complete knowledge' which has haunted western science for three centuries. Both in the hard sciences and in the so-called soft sciences, we have only a window knowledge of the world we wish to describe.

Introduction

The very notion of 'complexity' refers to a dual conception of nature: on one side we have 'simple' phenomena, such as the motion of a pendulum or the equation of state of a perfect gas; on the other, the 'complex' phenomena involved in life or human behaviour. An expected feature which emerges from the scientific history of the last decades is that this duality seems to vanish. Who would have expected that 'elementary' particles would have such an intricate structure; that the motion of an elastic pendulum would be comparable in complexity to that of fully developed turbulence? Moreover, the type of behaviour depends on external conditions. A fluid layer is described qualitatively using the image of molecular chaos. However, when we heat a fluid from below, molecular chaos is partly lost. Similarly, a chemical system, when driven far from equilibrium, develops space–time patterns which are again incompatible with the simple idea of merely chaotic motion of molecules.

All this has a deep effect on the way we look at the world around us. Obviously, it can no longer be symbolized by stable, periodic motion. Moreover, these unexpected discoveries have a drastic effect on our view of the relationship between 'hard' and 'soft' sciences. The traditional gap is narrow-

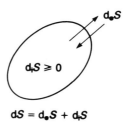

$$dS = d_\bullet S + d_+S$$

Fig.3.1. Second law of thermodynamics.

ing and we must again consider a transfer of knowledge from one field to another. This process has, in fact, already started in many disciplines, such as in the study of economics or insect societies. However, I must comment here on more basic aspects. I believe that the unifying feature in contemporary science is the 'rediscovery of time', which is an essential element both in recent developments in non-equilibrium physics and in the general mathematical theory of dynamical systems. We have to make more precise the road which leads from deterministic and time-reversible systems, such as those considered in classical dynamics, to systems involving stochastic and irreversible processes.

This chapter is in two parts: the first is devoted to phenomenological aspects. Here, the road goes through non-equilibrium thermodynamics. I believe that the most fascinating aspect is the study of the mechanisms which, in a sense, transfer irreversibility to space and matter. To illustrate this point, I shall consider the problems of life and of cosmology. The second part of the chapter will deal with the emergence of stochasticity and irreversibility in the field of classical dynamics. A good illustration is provided by cometary clouds.

Non-equilibrium thermodynamics and complex systems

Some 120 years ago (1865), Clausius formulated the second law of thermodynamics, introducing a new quantity: entropy. The fundamental importance of entropy is that as a result of irreversible, time-oriented processes, the entropy of our Universe (considered as an isolated system) is increasing. This was an amazing statement at a time dominated by classical mechanics.

Since Clausius, physics has dealt with two concepts of time: time as repetition and time as degradation. But it is obvious that we must overcome this duality. Neither repetition (the negation of time), nor decay (time seen as degradation), can do justice to the complexity of the physical world. We must therefore reach a third concept of time, which also contains positive, constructive aspects.

Let us first recall that a distinction may be made between two additive terms in the variation of entropy dS (Figure 3.1), d_eS and d_iS, respectively the exchanges (positive or negative) between a system and its environment, and the internal entropy production (always positive). For an isolated system, $dS =$

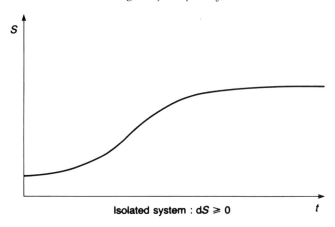

Fig.3.2. Monotonous increase of entropy for an isolated system.

d_iS is positive or zero (Figure 3.2). But what happens if the system is submitted to external constraints?

The first steps of thermodynamics were associated with the consideration of equilibrium states. There is no arrow of time for an isolated system at equilibrium. Later (1931), Onsager gave general relations for non-equilibrium thermodynamics in the near-equilibrium region: systems submitted to weak external constraints give linear responses to these constraints. These linear, reciprocal relations between fluxes and forces designated non-equilibrium thermodynamics as a worthwhile subject of study, since they extended the usual description of equilibrium systems – in terms of thermodynamic potentials whose extrema correspond to the final states – to a new description for non-equilibrium systems in terms of a new potential function: the entropy production.

Let us consider an example: a two-component system containing hydrogen and nitrogen is submitted to an external heat flow, maintaining a thermal gradient. As a result, one of the components, say H_2, accumulates in compartment I, the other in compartment II. In the frame of macroscopic thermodynamics, the entropy production per unit time can be written as the sum of two terms:

$$\frac{d_iS}{dt} = \underset{>0}{\text{heat flow}} + \underset{<0}{\text{diffusion}} \geqslant 0$$

The important point is that even in this simple case, we have a *coupled phenomenon*: heat flow is producing entropy, but diffusion goes *against* the concentration gradients (this is the so-called 'thermodiffusion' effect), and would lead to a negative entropy production *if acting alone*.

This shows that irreversibility generally has a *dual* aspect: it corresponds both to dissipation (here the heat flow) and to the formation of *order* (the thermal diffusion). That is why I have put on the cover of one of my books a

Fig.3.3. The Bénard hydrodynamical instability (as seen from above).
Liquid heated from below.

representation of a well-known Indian symbol, the dancing Shiva, who holds in one hand a fire which destroys, and in another a drum which creates!

Let us now turn to non-equilibrium thermodynamics valid for systems in far-from-equilibrium conditions. Systems submitted to strong constraints present responses that are no longer linear; accordingly, their behaviour displays a multiplicity of stationary states and a lot of new, related, dynamical character-istics such as bifurcations, hysteresis, and so on. Let us give two examples of this newly understood behaviour of non-equilibrium systems.

The so-called Bénard instability is a striking example of instability in a stationary state giving rise to a phenomenon of spontaneous self-organization; the instability is due to a vertical temperature gradient set up in a thin horizontal liquid layer. The lower face is maintained at a given temperature, higher than that of the upper face. As a result of these boundary conditions, a permanent heat flux is set up from bottom to top. For small temperature differences, heat can be conveyed by conduction, without any convection, but when the imposed temperature gradient reaches a threshold value, the stationary state (the fluid's state of 'rest') becomes unstable: convection arises, corresponding to the coherent motion of a huge number of molecules, and increasing the rate of heat transfer. In appropriate conditions, the convection produces a complex spatial organization in the system (Figure 3.3).

There is another way of looking at this phenomenon. Two elements are involved: heat flow and gravitation; under equilibrium conditions the force of gravitation has hardly any effect on a thin layer of about 10 mm. In contrast, far from equilibrium, gravitation gives rise to macroscopic structures. Non-equilibrium matter becomes much more sensitive to the outer world conditions than matter at equilibrium. I like to say that at equilibrium, matter is blind; far from equilibrium it may begin to 'see'.

Consider, secondly, the example of chemical oscillations. Ideally speaking, we have a chemical reaction whose state we control through the appropriate injection of chemical products and the elimination of waste products. Suppose that two of the components are formed, respectively, by red and blue molecules in comparable quantities. We would expect to observe some kind of blurred colour with perhaps some occasional flashes of red or blue spots. This is, however, not what actually happens. For a whole class of such chemical reactions, we see the whole vessel becoming red, then blue, then red again in sequence: we have a 'chemical clock'. In a sense, this violates all our intuitions about chemical reactions.

We used to speak of chemical reactions as being produced by molecules moving in a disordered fashion and colliding at random. But, in order to synchronize their periodic change, the molecules must be able to 'communicate'. In other words, we are dealing here with new supermolecular scales – in both time and space – produced by chemical activity.

The basic conditions to be satisfied for such chemical oscillations to occur are auto- or cross-catalytic relations, leading to 'non-linear' behaviour, such as are described in numerous studies of modern biochemistry. Remember that nucleic acids produce proteins, which in turn lead to the formation of nucleic acids; there is a cross-catalytic loop involving proteins and nucleic acids.

Non-linearity and far-from-equilibrium situations are closely related; their effect is that they lead to a multiplicity of stable states (in contrast to near-equilibrium situations, where we find only one stable state). This multiplicity can be seen on a 'bifurcation diagram' (Figure 3.4): we have plotted the solution of the problem X against some bifurcation parameter λ (X would be, for example, the concentration in some chemical component, and λ could be related to the time the molecules spend in the chemical reactor). For some critical value of the control parameter, say λ_c, new solutions emerge. Moreover, near the bifurcation point, the system has a 'choice' between the branches: we could therefore expect a stochastic behaviour; near a bifurcation point, fluctuations play an important role.

The fact that dissipative systems present a behaviour which can be described in terms of attractors implies a basic difference with mechanics. Suppose we have some foreign celestial body approaching the Earth. This would lead to a deformation of the Earth's trajectory, which would remain for astronomical times: mechanical systems have no way to *forget* perturbations. This is no longer the case when we include dissipation. A damped pendulum will reach a position of equilibrium, whatever the initial perturbation from it.

We can also understand in quite general terms what happens when we drive a system far from equilibrium. The 'attractor' which dominated the behaviour

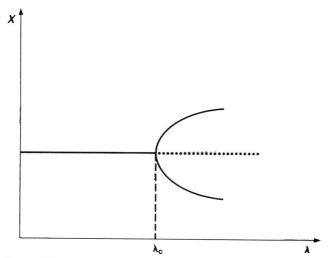

Fig.3.4. Bifurcation between stationary branches for a control parameter λ.

of the system near equilibrium may become unstable, as a result of the flow of
matter and energy which we direct at the system. Non-equilibrium becomes a
source of order; new types of attractors, more complicated ones, may appear,
and give the system remarkable new space–time properties.

We have seen that dissipative systems may forget perturbations: these
systems are characterized by attractors. The most elementary attractors are
points or lines, as represented in Figures 3.5 and 3.6. In Figure 3.5 we have a
point attractor in a two-dimensional space (X_1 and X_2 may be concentrations of
some species): whatever the initial conditions, the system will necessarily
evolve towards the point attractor. In Figure 3.6, we have a line-attractor:
whatever the initial conditions, the system will eventually evolve along this
line, called a limit-cycle.

But attractors may present a more complex structure; they may be formed
from a scattered set of points (Figure 3.7), whose distribution may be dense
enough to permit us to ascribe them an effective non-zero dimensionality. For
example, the dimension of the attractor in Figure 3.7 may be any real number
between 2 and 3. Following the terminology of Benoit Mandelbrot, one may
call this a 'fractal' attractor. The behaviour of systems in the presence of such an
attractor can be described using the new qualitative non-linear dynamics. For
example, one can see that trajectories on the chaotic attractor in Figure 3.8 are
converging along one direction, diverging along the other; these characteristics
are described by the respective Lyapunov exponents attached to these
directions.

Such systems have unique properties reminiscent of, for instance, turbulence which we encounter in everyday experience. They combine both
fluctuations and stability. The system is driven to the attractor; still, as the latter
is formed by so 'many' points we may expect large fluctuations. One often
speaks of 'attracting chaos'. These large fluctuations are connected to a great

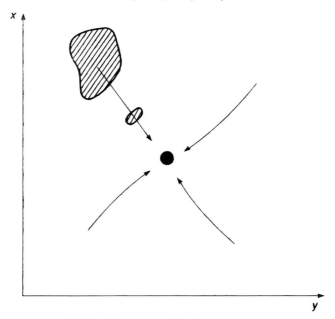

Fig.3.5. Point attractor in phase space (x, y).

sensitivity with respect to initial conditions. The distance between neighbour-ing trajectories grows exponentially in time (this growth is characterized by the so-called Lyapunov experiments). Attracting chaos has now been observed in a series of situations, including chemical systems or hydrodynamics; but the importance of these new concepts goes far beyond physics and chemistry proper. Let us indicate some examples studied recently.

We know that climate has fluctuated violently over the past. Climatic conditions that prevailed during the last two or three hundred million years were extremely different from what they are at present. During these periods, with the exception of the quaternary era (which began about two million years

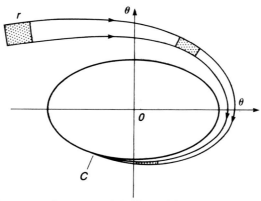

Fig.3.6. Limit-cycle attractor. As in Figure 3.5, one sees a contraction of the area associated to the convergence of trajectories to the limit cycle.

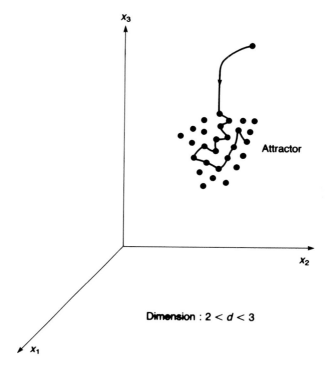

Fig.3.7. 'Fractal attractor' formed by a multiplicity of points with a 'dimension' between 2 and 3.

ago), there was practically no ice on the continents, and the sea level was higher than at present by about 80 metres. A striking feature of the quaternary era is the appearance of a series of glaciations, with an average periodicity of one hundred thousand years, on which is superimposed an important amount of 'noise'. What is the source of these violent fluctuations (Figure 3.9) which

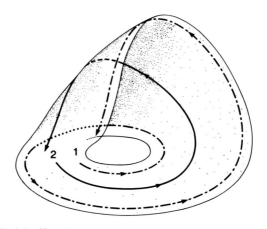

Fig.3.8. Chaotic attractor corresponding to a Rössler model.

[76]

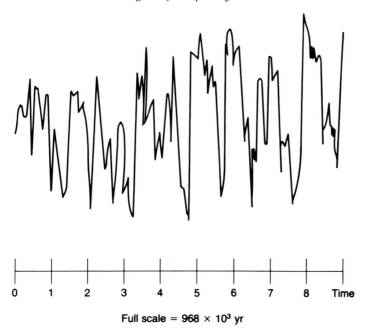

Full scale = 968 × 10³ yr

Fig.3.9. Full scale = 968 × 10³ yr. Series of temperatures for the Earth's
climate over a million years.

have obviously played an important role in our history? There is no indication
that variation in the solar radiation may be responsible for them.

A recent analysis by C. and G. Nicolis has shown that these fluctuations can
be modelled in terms of four independent variables which form a non-linear
dynamical system leading to a chaotic attractor of dimension 3.1 embedded in a
phase space of dimension 4. The variability of climate could have been thought
of as resulting from the interplay of a large number of variables acting in a
deterministic fashion; the situation would then be very similar to the outcome
of the law of large numbers. The new insight is that this is not so. The temporal
complexity can be described by only four independent variables. We may
therefore speak of an intrinsic complexity or unpredictability of climate.

In a quite different field, recent work has shown that the electrical activity of
the brain during deep sleep, as monitored by electroencephalogram (EEG),
may be modelled by a fractal attractor. Deep sleep EEG may be described by a
dynamics involving five variables; again, this is very remarkable in that it
shows the brain acting as a system possessing intrinsic complexity and
unpredictability.

It is this instability which permits the amplification of inputs related to
sensory impression in the waking state. Obviously, the dynamical complexity
of the human brain cannot be an accident. It must have been selected for its
very instability. Is biological evolution the history of dynamical instability,
which would be the basic ingredient of the creativity characteristic of human
existence?

[77]

Ilya Prigogine

Irreversibility and information in biology and cosmology

Let us now present some (necessarily rather speculative) remarks about two problems which have most excited the imagination of man: the origin of life and the origin of the Universe.

We wish to emphasize that irreversible processes have likely played an essential role in inscribing, so to speak, time into matter at an early stage of the Universe and producing the information carried by basic biochemical compounds such as DNA.

Indeed, a basic feature of biomolecules is that they are carriers of information. This information is quite similar to a text, which has to be read in one direction, exactly like a text in human language. In this sense, biomolecules present a broken symmetry. It is natural to relate this broken symmetry to the broken time symmetry which is expressed in the second law of thermodynamics. Already, in the simple case of the Bénard instability, irreversibility is transformed into a pattern. However, the Bénard flow pattern persists only as long as heat crosses the liquid layer. In contrast, life has an extraordinary degree of persistence, as it originated 3.4 billion years ago.

One of the basic questions in the origin of life is a better understanding of the selection of nucleotide sequences. Suppose we take a protein 100 amino acids long. It is well known that in nature there exist $N = 20$ kinds of amino acids. This would lead to 20^{100} possible sequences. Non-equilibrium conditions are essential to reduce this number. We have to produce sequences to which we

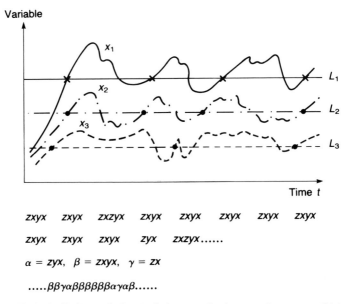

Fig.3.10. Coding of chaotic behaviour leading to 'incompressible' information (from Nicolas, G. and Subba Rao, G. and J., private communication.)

can associate some 'information'. This excludes purely random sequences, as well as periodic sequences which could be compressed into a few symbols. To borrow a term from communication theory, we need *incompressible* information.

It has recently been shown that chaotic chemical reactions provide us with a very simple model for this inscription of irreversibility into matter. As suggested by G. Nicolis and G. and J. Subba Rao, we may consider a chemical reaction involving three components: X_1, X_2, and X_3. At each crossing of some threshold by one of these concentrations, we assume that the molecule of the corresponding component is added to some polymer (Figure 3.10). As a result, we obtain a sequence which indeed carries an incompressible information.

Of course, with a limit-cycle reaction, the analogous sequence would carry a compressible information, as in this case we could only expect periodic strings like *xyzxyzxyz*. It has been shown by Nicolis and Subba Rao that the sequence obtained from the chaotic reaction, when read in one direction, corresponds to a Markov chain. Irreversibility leads here to a 'text'. This is a simplified model, but one may hope that real-world experiments can be performed to produce polymer chains in far-from-equilibrium conditions, separated from the thermodynamical branch by bifurcation points. Here, we may expect to find information-carrying molecules which could indeed be plausible ancestors for the biomolecules of today. The important point we want to emphasize here is that non-equilibrium processes with chaotic dynamics can indeed generate 'information'.

We have just made some remarks on the problem of the origin of life but, obviously, the most striking of all such problems is the one of the origin of our Universe. The modern history of cosmology has been a dramatic one. As is well known, in 1917 Einstein proposed a static model of the Universe. General relativity seemed to him to be the ultimate achievement, as it gave the unification of gravitation with space–time. But this static image was to be rapidly given up, when in 1922 Friedmann showed Einstein's equations were unstable, and Hubble's observations pointed in the direction of an evolving, expanding Universe. Later, the celebrated residual black-body radiation suggested that beyond this geometrical evolution there is a more basic thermal evolution.

We thus come to the modern 'standard' cosmology (Figure 3.11). It presents two fundamental aspects: the temperature of the Universe increases mono-tonically when we come close to the 'big bang', while, according to the adiabatic type of evolution, the entropy of the Universe remains constant. We cannot go into the details here, but two obvious questions are implied: how could our Universe start from this high temperature; and how could entropy reach this level at once? It is very tempting to reconsider these questions from the viewpoint of irreversible processes. In the Universe we find a remarkable duality: it is principally made up of photons and baryons. A remarkable insight into the structure of the Universe is given by the so-called 'specific entropy', S, defined as the ratio of the number densities of photons to baryons. The usually accepted value is $S \simeq 10^8 \sim 10^9$.

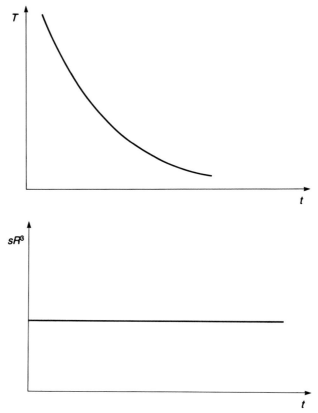

Fig.3.11. Initial evolution assumed in the standard cosmology. Temperature goes to ∞ for $t \to 0$ and entropy remains constant.

Today, baryons are considered to be non-equilibrium objects which are either in a metastable state or decaying, while the photons are 'waste products' which can no longer decay into other known forms of matter. So, most of our Universe is at thermal equilibrium from this 'particle' point of view; but (fortunately for us) there is a small non-equilibrium residuum. Such a duality is typical for coupled non-equilibrium processes. On Earth, solar radiation gives rise to positive entropy production, most of which leads to dissipation in the form of photons: in a schematic way,

$$\frac{d_i S}{dt} = \underset{>0}{\textit{production of photons}} + \underset{<0}{\textit{production of biomolecules}} \geqslant 0.$$

This reminds us of the situation mentioned on page 71 (effect of thermal diffusion), where the entropy production is also the sum of a positive and negative term. If in the example of thermal diffusion the heat flow is discontinued, the concentration of the components in the two compartments will slowly equalize.

The important point is that the part of the system which is slowly evolving to equilibrium has itself been brought out of equilibrium by a *non-equilibrium process*. Could it be that matter (the baryons and leptons) was the out-of-equilibrium product of some cosmic non-equilibrium process which has also produced the black-body radiation?

There is also a basic duality which appears in general relativity: on one side, space–time; on the other, matter. These two aspects are connected through Einstein's field equations. In recent years, it has become quite popular to describe the creation of the Universe as a 'free lunch'. The energy of the vacuum would be transformed into a *positive* energy (the 'matter'), and a *negative* energy (the gravitational field energy)

However, if matter is indeed created from the vacuum, this should be an *irreversible, entropy-producing process*. In this view, the entropy of the vacuum would be zero. Matter would be the 'contamination' of space–time, carrying the entropy generated by this creation process.

As the lecture on which this chapter is based was given at Cambridge, a well-known centre for the study of cosmology, I could not resist the pleasure of giving some further details. Gehéniau and I have tried, with the use of a well-known formalism (the pseudotensor), to express this 'irreversible' transfer of 'gravitational' energy to matter. We used *conformal* coordinates; i.e., the metric represented by

$$(ds)^2 = F^2(ds)^2$$

where $(ds)^2$ is the usual Minkowski metric appropriate for flat space–time and F the conformal factor. Let us limit our discussion to the open universe ($k = -1$). Then F starts at zero and reaches $F = 1$ asymptotically (in fact, on a short timescale of a fraction of a second). It is during this sort of 'phase transition' from $F = 0 \rightarrow F = 1$ that matter will be created. Moreover, the creation of matter leads to a redefinition of pressure p and density ρ:

$$\vartheta = F^3 \vartheta, \quad \vartheta = F^3 p$$

This leads to the fascinating possibility of cosmological models without matter singularity. Indeed, while $\vartheta \rightarrow \infty$ for $t \rightarrow 0$ (typically, as $1/R^3$ where R is the Friedman radius), $F^3 \rho$ would go to a constant for $t \rightarrow 0$. The Universe would start cold and empty and develop its matter and energy content as processes coupled to the evolution of space–time. The Universe would present us with the most grandiose example of self-organization due to irreversible processes (Figure 3.12).

This method suffers from the fact that it is not covariant (as the result of using pseudotensors), but we can put it into a covariant form by including a term

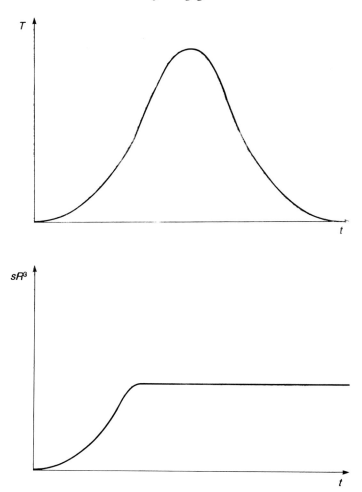

Fig.3.12. Possible initial behaviour for the temperature of the Universe and its entropy (from Géhéniau, J. and Prigogine, I., 'Foundations of Physics', *Proc. Natl. Acad. Sci.*, 1986).

corresponding to matter creation in Einstein's field equation.[†] Obviously, the additional term will be *formally* analogous to that introduced in the old days of the steady state theory. However, here it is essentially related to the irreversible entropy production due to matter creation, only in one limited period, and leads therefore to an evolutionary (and *not* steady state) Universe.

In special relativity, we have the well-known equivalence principle between energy and matter. Here, there appears to be something like a new equivalence principle between *matter* and *entropy*, as each creation of matter corresponds to an entropy increase. The zero of entropy is no longer the crystal at 0 K, as in

[†] Einstein's field equations described the interaction between matter and gravitational field as reversible; whatever type of irreversible processes which may have existed in the early Universe, they will require a generalization of Einstein's field equations.

conventional thermodynamics, but the vacuum out of which matter emerges. In this perspective the very idea of 'elementary' particles is gone; particles carry only more or less entropy, more or less information. Obviously, to make sense, such a theory presupposes a time symmetry breaking in the fields leading to the coupling between matter and gravitation, and here we come to the dynamical aspects of irreversibility.

From instability to irreversiblity

What do we mean by laws of Nature? Let us take the example of coin-tossing, already discussed by Poincaré. Let H denote heads and T stand for tails. The familiar outcome is traditionally represented by equal probabilities: $P_H = 0.5$ and $P_T = 0.5$. Does this outcome correspond to a fundamental law of Nature? If we assume that a deterministic representation of coin tossing (Newton's laws of motion) is possible, should not the outcome be 0 or 1, according to the initial conditions?

We can now analyse this situation in more detail. Suppose we restrict the initial conditions to smaller and smaller subsets of the initial conditions set. If, starting with some initial condition determined with finite precision, the outcome corresponded to the deterministic 0/1 prediction, the basic law of coin tossing would indeed be deterministic. However, if for arbitrary initial conditions, whatever their (finite) precision, the outcome remains 0.5/0.5, then the law of coin tossing is a statistical law; we are then to accept that the outcome of a single coin toss is unpredictable – but that we can make statistical predictions if we repeat the game.

What is the difference between the two situations? Let us represent the initial conditions on a segment of a line. Two things can happen:

> Either there are parts of this segment on which all points (with some appropriate measure) lead to H, or to T. Then we come to a deterministic situation. The dynamical system is then *stable* (sufficiently small perturbations will not change the outcome).
>
> In contrast, if each point leading to H is surrounded by points leading to T (and vice-versa), we can only make statistical predictions. The system is then *unstable*. Of course, this issue would change if we accepted the idea of infinite precision.

It is reported that Bohr stated to Planck that in quantum mechanics, coordinates and momenta cannot be determined simultaneously. Planck answered: 'But God knows both!', to which Bohr allegedly remarked that physics deals only with what *man* can know.

The illusion of complete (infinite) knowledge comes, it seems, from the historical fact that classical science started with the study of periodic motions. The return of the Sun and the regularity of celestial phenomena have deeply influenced man's thought since palaeolithic times. It led to the prototype of

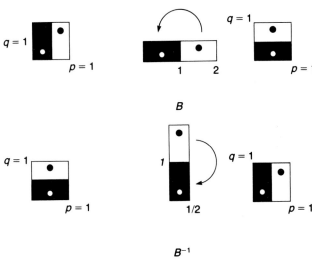

Fig.3.13. The baker transformation in phase space (p, q).

knowledge as expressed in classical physics. But this regularity of periodic processes is not the general case. The message conveyed by the second principle of thermodynamics is that we are not living in a world that can be described in terms of periodic motions. It is an unstable world, which we know through a 'finite window'.

Let us consider a highly unstable dynamical system and show how this instability leads to irreversibility. The system we shall consider corresponds to the so-called 'baker transformation'. Here the phase space is a square on which we apply the operations well known by bakers: flatten, cut, fold (Figure 3.13). In this transformation the surface area is conserved, but points are scattered over the whole phase space.

The baker transformation induces a particular geometric structure on the square. It is clear that points separated by a vertical (horizontal) segment will find their distance reduced (magnified) by a factor 2 at each iteration. This determines two foliations of the phase space, one into horizontal (dilating) fibres, and one into vertical (contracting) fibres (Figure 3.14). Each point is at the intersection of a contracting and a dilating fibre. All points on the same contracting fibre tend to the same phase point (rather, trajectory) for $t \to +\infty$. They have the same (asymptotic) *future*. All systems on the same dilating fibre have the same past (for $t \to -\infty$). The essential point is that fibres characteristic for unstable dynamical systems have a *broken time symmetry*. The dilating fibres are time-reversal conjugates of the contracting fibres.

For this reason, all finite initial data require a non-local description with broken time-symmetry. Whatever the precision offered, we shall never be able to distinguish points sufficiently close on the *contracting* fibres. We must therefore replace a point by an ensemble of points on the same contracting fibre. This leads to the consideration of bundles of trajectories (a non-local

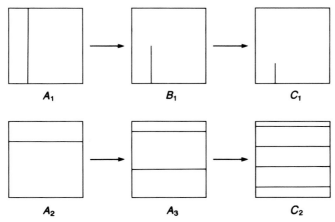

Fig.3.14. Contracting and dilating fibres for the baker transformation.

concept) with broken symmetry as this construction is based essentially on the distinction between contracting and dilating fibres. Here we cannot go into the details, which are of a rather technical nature. But in order to avoid misunderstandings, let us emphasize, however, that the limitation to finite precision does not introduce any subjective feature. It would indeed be difficult to accept that our measurements could be based on mechanisms other than natural processes: any interaction between dynamical systems also involves a limited 'window' formed by their limited sensitivity, exactly as in our observations. That makes the description of our world inherently irreversible and stochastic, even in the field of classical dynamics.

Celestial dynamics – cometary clouds

Dynamical systems such as the baker transformation are of course highly idealized. It is therefore interesting to describe an example in celestial dynamics in which irreversibility and stochasticity also play an essential role. For a long time celestial dynamics appeared to be the stronghold of a deterministic and time-reversible description of nature. It is therefore important to see how the new concepts now diffuse into the very field which gave rise to the paradigm of classical physics.

The example to which I shall refer is the capture of comets in the Solar System through the effect of Jupiter–Sun coupling. In its simplest version, it is an example of the celebrated three-body problem (here we have the Sun, Jupiter and the comet). We may even consider the *restricted* three-body problem corresponding to planar motion.

The difficulty is that the capturing process occurs through a strong resonance effect between the period of Jupiter and the motion of the comet. This leads to the difficulty of the 'small denominator', so central since the classical work of Poincaré. This has been solved to some extent through the work of Kolmogorov, Moser and Arnold, but this theory does not apply to the case of

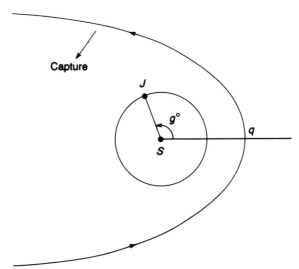

Fig.3.15. Orbit of a comet in the Jupiter–Sun gravitational field.

strong resonance, such as that occuring for the capture of comets on *near-parabolic motion.*

However, we have shown some years ago how some strong resonance effects can be taken into account through analytical continuation. The theory of our celestial system then becomes very similar to kinetic theory, as the resonances lead to *irreversible* transfer of energy. Curiously, non-equilibrium thermodynamics starts with the three-body problem! (See Figure 3.15.)

Small transfers of energy lead from parabolic motion into either elliptic motion (when energy is transferred from the comet to Jupiter) or to hyperbolic motion (in the opposite case). With a given flow of comets coming in, we may observe the formation of cometary clouds containing comets, which make a

Table 3.1. *The number of revolutions of a comet before escaping the solar system, as calculated with different accuracies. Initially $P_0 = 0$. The first column shows the initial phase angles g^0 of Jupiter in degrees. The second to last columns show the number of revolutions calculated with the accuracy of six, seven, . . ., up to 16 digits. The number 1000 in the table means that the number of revolutions was equal to or larger than 1000*

g^0	1E−6	−7	−8	−9	−10	−11	−12	−13	−14	−15	−16
162.0	757	38	236	44	12	157	7	8	1000	143	17
165.6	2	85	6	4	13	5	3	3	4	6	3
169.2	3	11	2	3	97	39	22	417	11	4	3
172.8	7	458	10	1000	5	8	1000	5	636	1000	23
176.4	23	21	938	39	40	14	8	31	1000	16	136
180.0	8	36	383	19	4	64	1000	23	474	73	27
183.6	3	118	2	1000	400	4	1000	7	32	116	6
187.2	2	42	4	4	1000	6	8	4	18	3	3
190.8	11	3	2	5	155	5	7	20	1000	16	6

number of revolutions around the Sun–Jupiter system, and then acquire a positive energy and leave the cloud. We may calculate the 'lifetime' of a comet (i.e., the number of revolutions it will make in the cloud before leaving it) either by studying the motion of a single comet or by considering the dynamics of ensembles of comets.

The remarkable result obtained by Petrosky and Broucke is that the lifetime of a comet, calculated for a single comet, is an ill-defined concept (in spite of the fact that initial conditions are given). This is illustrated in Table 3.1. The lifetime fluctuates violently, whatever the numerical process imposed, exactly as the outcome of a game of chance. It is only at the level of an *ensemble* of comets that we can obtain reliable predictions on the structure of cometary clouds. This example, as well as the study of the Kirkwood gap in the asteroid belt, where the situation is somewhat similar, shows that irreversibility and stochasticity now enter into celestial dynamics. The classical paradigm of dynamics, stressing determinism and reversibility, came from an undue generalization of the behaviour of simple systems such as that involved in two-body problems.

Concluding remarks

The recognition that complexity plays an essential role in all levels of our description of nature leads us to reconsider the relation between states and laws, between 'being' and 'becoming'.

In the classical view (including quantum mechanics), states are time symmetrical and propagated by laws preserving time symmetry, such as the unitary transformations of physics. To take irreversibility and stochasticity into account, we have to consider states with broken time symmetry, propagated by laws which are themselves due to a broken time symmetry, such as the second law of thermodynamics. Here we deal with the transition between two mathematical formalisms, leading from groups into semigroups.

We come to a new level of understanding in which rationality is no longer identified with 'certainty', nor probability with 'ignorance'.

Al. Koyré has aptly expressed the context of the scientific revolution of the seventeenth century in the title of his book, *From the Closed World to the Infinite Universe*. This infinite universe was the universe of Newton, infinite in both space and time. However, this infinite universe remained a static one for nearly three centuries. It is only in the last decades that we are gradually acquiring the tools necessary to describe a different type of irreversible, entropy-generating process. Are we at the beginning of the second stage in the scientific revolution?

Acknowledgements

I would like to mention the helpful comments and suggestions of J. Giehéniau, B. Misra, G. Nicolis and T. Petrosky. During the preparation of this paper, I received the support of the Robert A. Welch Foundation (Houston, Texas), and the Institute for Constructive Capitalism (Austin).

Ilya Prigogine

Further reading

Gribbin, J. *In Search of the Big Bang.* Bantam, New York, 1986.

Mandelbrot, B. *The Fractal Geometry of Nature.* Freeman, San Francisco, 1983.

Nicolis, G., and Prigogine, I. *Self-Organization in Nonequilibrium Systems.* Wiley, New York, 1977.

Nicolis, G., and Prigogine, I. *Exploring Complexity.* Freeman, San Francisco, 1986; and Piper Verlag, München, 1986.

Prigogine, I. *From Being to Becoming.* Freeman, San Francisco, 1979.

Prigogine, I., and Stengers, I. *Order Out of Chaos.* Bantam, New York, 1984.

[4]

Human origins and evolution

David Pilbeam

Introduction

Palaeoanthropology, the study of human evolution, is fascinating and difficult. It is fascinating because it concerns humans, and because answering relevant questions requires knowledge of a broad range of data and approaches: fossils, ethological studies of living primates, hunter–gatherer studies, comparative genetics, archaeology, sociobiology. It is difficult because it involves the analysis and explanation of unique events, and unique events have unique causes. It is also difficult because it involves us.

Paleoanthropologists have reached a reasonable consensus on most of the stages of human evolution: how many stages (species or groups of species), their timing, their documentation. But more interesting questions about the behaviour of past species, and about the dynamics of change, are still quite heatedly debated. This is only serious fun if we can spell out why there are disagreements, what can be done about them, and how to proceed on what are at present the more intractable problems. There has been a significant 'non-scientific' component to ideas about human evolution, so understanding their genesis is itself important. Retelling the history of ideas in any subject can be dull, but much of today's research agenda still bears the imprint of questions that used to be important, so some 'historiography' will occasionally surface here.

Palaeoanthropologists would like to answer the question: why are humans so peculiarly and successfully (or so we like to think) different from their closest relatives, and how did they get to be that way? The first step towards answering the question is to specify the ways in which humans differ from non-humans. Since a list of specifically human features could be extremely long and because some are more important than others, this involves being selective. Selectivity clearly opens up another problem: what one selects will determine the parts of the past one wishes to reconstruct. All accounts of human evolution start, then, with a selection of differences between humans and non-humans. Essentially, all the early accounts used a difference as the cause of the initial (presumed fundamentally important) step taken by the hominids. (Hominids is the colloquial version of *Hominidae*, the formal name for the group containing humans, their ancestors and relatives, separate from our closest ape relative.) Keith Thomas has provided an informative and amusing review of 'human uniqueness' since Aristotle, and a number of palaeoanthropologists have discussed this topic. One of my favourite anatomical differences was eloquently expressed by an early Stuart doctor:

> Man is of a far different structure in his guts from ravenous creatures as dogs, wolves, etc., who, minding only their belly, have their guts descending almost straight down from their ventricle or stomach to the fundament: whereas in this noble microcosm man, there are in these intestinal parts many anfractious circumvolutions, windings and turnings, whereby longer retention of his food being procured, he might so much the better attend upon sublime speculations, and profitable employments in Church and Commonwealth.

The range of favourite human peculiarities has covered just about every conceivable human feature from feeding to brains to sex, and as the fossil record and theoretical approaches have changed, so too has the list. The main problem has been that most accounts of human evolution have been relatively simple – two- or at most three-stage stories – so that if a modern human difference is taken as the cause of the initial split, almost inevitably, early hominids will seem more human than they probably were. This is a particular problem with intangibles such as social behaviour. Because early hominids generally ended up looking rather human in many earlier accounts, arguments among palaeoanthropologists have often been more about relationships between hominid species than about reconstructions of behaviour.

Given the history of palaeoanthropology, this is, at least in part, understandable. Scientists writing about human evolution a century ago had essentially no fossil record and were most concerned with demonstrating that humans had indeed evolved. Explanations for why humans evolved usually invoked the self-evident success of human characteristics such as bipedalism or tool use and rarely went through any elaborate steps of reconstructing past hominids. Indeed, such assumptions linger today. How often have you asked yourself how long after humans went extinct would it take before some other animal evolved recognizably human traits? If it seems obvious why evolutionary changes occurred, if our ancestors seem to have resembled us, then adequate reconstruction of those ancestors takes on a low priority.

The hominid fossil record grew steadily through the last portion of the nineteenth century and the first half of the twentieth, but accounts remained simple two-stage and one-step affairs and it remained self-evidently obvious why human features evolved: because they were human. Throughout this first half of the twentieth century it had seemed clear that, just as a large brain was perhaps the most conspicuous human distinction, the brain had led the way in human evolution. This made it difficult to accept the bipedal but small-brained australopithecines (genus *Australopithecus*) as hominids when these were discovered during the 1920s and 1930s. However, the recovery in 1947 at

Sterkfontein in South Africa of a partial skeleton showed that *Australopithecus* had an ape-sized brain attached to a bipedal and distinctly human-like body. Rapidly, the australopithecines became accepted as the earliest hominids. Yet almost as rapidly their behaviour came to be seen as quite like that of humans; the distinction between 'hominid' and 'human' was blurred. Clifford Jolly proposed an explicitly two-phase evolutionary scheme in which the adaptive explanations for early hominid (australopithecine) anatomy explicitly differed from that for later, larger-brained *Homo*, but there has been a tendency for earlier hominid behavioural reconstructions to continue 'drifting' in a modern direction.

There has, however, been a fascinating recent swing of the palaeoanthropo-logical pendulum: each stage in the human story is now seen as significantly less like modern humans than was thought, even quite recently. The extent to which this is just a swing rather than a real change based on new data and solid interpretation remains to be seen.

There is now general agreement on the most interesting and important human features whose trajectory we would like to track. These are listed in Table 4.1. There are many odd things about humans, but most obvious among them are being upright and bipedal which frees our hands; tool-making; a brain three times the size of those of apes, capable of complex and mostly learned cultural behaviour based on symbols – particularly speech; marriage and kinship networks, and peculiar patterns of child rearing (with paternal involvement) and sexual behaviour; use of both hunted-animal and collected plant foods, and how we process and share them; our odd use of space, dispersing from a regular home base and returning to it after the daily round; making home bases the centres for many other activities, our ability to do this being based in our degree of environmental control – flowing in particular from the ability to control fire.

This list specifies the agenda – those features which must be explained. To show how they evolved we need to be able to do three things. First, describe the extinct hominid species which are now our close ancestors and cousins. By this I mean describe them qualitatively or, where appropriate, quantitatively, in the sense of a state description involving the selection of critical state variables. This involves, among other things, finding more fossils; even well-documented stages could stand improvement, while some are essentially or

Table 4.1. *Characteristic human features*

1. Usually non-forested habitats
2. Relatively very large brain
3. Slow maturation; female ovulation concealed, no obvious oestrous
4. Omnivorous; food being hunted, gathered, transported, shared, prepared, stored
5. Bipedal; central place foraging
6. Complex cultural behaviour, including language
7. Social organization built around marriage; prolonged infant care
8. Tool–making, technological skill, dependence on equipment

Table 4.2. *List of critical behavioural–ecological attributes*

1. Ecological variables
2. Body size and dimorphism
3. Brain size
4. Life history patterns
5. Diet and subsistence behaviour
6. Positional and ranging behaviour
7. Male–male relationships
8. Female–female relationships
9. Male–female relationships
10. Adult–infant relationships
11. Intergroup relationships
12. Communicative behaviour
13. 'Technological' behaviour

completely unsampled. Second, the relationships of these species must be analysed and their stratigraphic ranges then used to establish a sequence of ancestors and descendants. Third, each stage in the sequence, species of group or species, must be reconstructed behaviourally, as much as possible as though it were alive today (what I call the 'Time Machine' approach). This involves pretending that past species, their communities and their environments, can be studied like living ones; we must approach the behavioural reconstruction of past species by asking the same questions we would ask if we were trying to draw a total picture of, for examples, elephants or chimpanzees or wild dogs or human foragers. Table 4.2 lists some of the critical features we would want to know. It is very important to remember that not all these important questions can be answered now, and some will never be answered. But we must start with the questions, not with the answers we can get only from a traditional palaeontological approach to fossils.

Studies of the world today, including attempts at understanding how it is organized, are essential to any interpretation and understanding of the past. We must understand the principles that inter-relate ecology, behaviour, and anatomy in the present before we can bring the past alive, because the past consists only of anatomy and archaeology and ecological possibilities and constraints. We are fortunate indeed that sociobiological theory has been so productive and that studies of animal and plant communities have been so lively – especially those of our closest living relatives, the apes.

The main stages of human evolution are listed in Table 4.3. The major steps were the achievement of a kind of bipedalism, perhaps with the origins of hominids around 5 my or more ago; the diversification of the small-brained bipedal australopithecines between 5 my and 1.5 my ago; the shift to stone flake-using and increased brain size more than 2 my ago with the origin of *Homo* from an australopithecine, and the addition of more meat to the diet; the development of a distinct, omnivorous, scavenging and/or hunting phase, based on patterned stone tools and probably speech, which was non-human but non-ape in behaviour pattern; the evolution of fully modern humans by 40 ty ago, probably considerably more, perhaps correlated with the appearance of fully modern language capacities; finally, and as a consequence of this last step, a very recent shift beginning 10 ty ago from food gathering to food production, which led ultimately to the formation of States and what we like to call 'Civilization'.

The most interesting steps or transitions in hominid evolution – and transitions are inherently difficult to study – are the origin of hominids, the origin of *Homo*, and the origin of modern *Homo sapiens*. To begin to understand these transitions we need an adequate description of both antecedent and descendent states or stages. The remainder of this chapter is devoted to a brief, incomplete, and critical review of current evidence and interpretation.

The hominoid radiation and hominid origins

For this discussion of hominid origins I shall briefly trace the odyssey of the early hominoid *Ramapithecus*. The first piece of *Ramapithecus* to be recognized was a fragment of upper jaw from India, found in 1932, and now dated to between 7 and 8 my ago. Beginning in the early 1960s Elwyn Simons, and later Simons and I, wrote a series of papers arguing that *Ramapithecus* was similar to the undoubted hominid *Australopithecus* in jaws and teeth, the only common parts then known. These similarities included small canines, large cheek teeth, thick enamel caps on the teeth, and robust jaws. We believed that these were special, evolved, or 'derived' features shared exclusively by these genera, and indicating a specific evolutionary link between them. In our view, this made *Ramapithecus* an early hominid, and its small canines suggested to us, following a much older line of argument, that it might have been a tool-user and even possibly something of a biped. *Ramapithecus* fossils were thought to go back 14 my or more, and the base of the human family was therefore extended to at least 14 my ago (Figure 4.1).

But in the 1960s another totally different track was being followed. Some biochemists and genetically oriented anthropologists, including Morris

Table 4.3. *Principal stages of human evolution*

Late Miocene 7–5 my: Divergence of human and chimpanzee lineages, establishment of significant bipedalism

Pliocene 5–1.6 my: Diversification and evolution of bipedal, small-brained, open-country, ape-like australopithecines

Late Pliocene 2.5–2 my: Appearance of *Homo*: large-brained, stone-flaking, more omnivorous, non-apelike but non-human behaviour patterns

Late Pleistocene from 125 ty: Evolution of modern *Homo sapiens*, anatomically and probably behaviourally similar to living humans

Latest Pleistocene from 10 ty: Domestication of plant and animal resources to the extent that populations may increase in density and become sedentary, eventually leading to urbanism and States

[93]

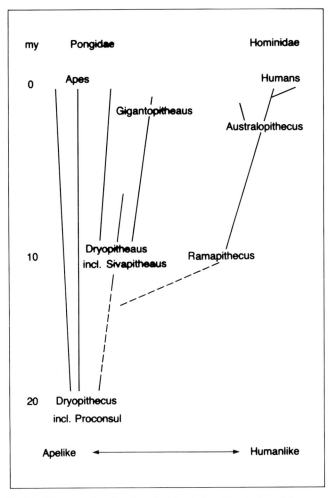

Fig.4.1. Diagram showing hominoid relationships believed by some palaeoanthropologists in the 1960s and 1970s.

Goodman, Allan Wilson, and Vincent Sarich, began to study the comparative genetics of living species. Initially working at the level of amino-acid sequences and antigenic reactivity of proteins, later moving to sequencing and other methods of comparing DNA, these studies have been among the most exciting in recent biology. To almost universal suprise, in virtually every genetic system studied, humans and African apes (common chimpanzee, bonobo or pygmy chimpanzee, and gorilla) are very similar, while the Asian apes, orangutan and gibbon, are different. We are, in a *genetical* sense, African apes. This was surprising because in the preceding half-century a consensus had arisen that the large apes (chimps, gorilla, orang) formed a monophyletic group – closely related species sharing a common ancestor – while hominids were more distant. Darwin, Huxley, and Haeckel were thus proved right in their earlier belief that humans were closest to African apes.

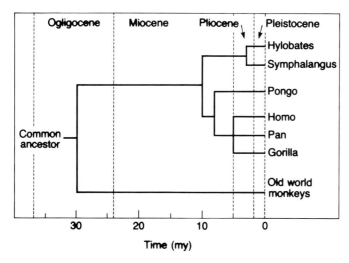

Fig.4.2. Evolutionary tree of hominoids and Old World Monkeys derived from albumin comparisons, based on a hominoid–monkey split of 30 my.

This was a surprising discovery, and not accepted without a fight, although it is widely viewed as correct today. More surprising and much harder to accept was Sarich and Wilson's claim that single proteins could be used as 'molecular clocks' which evolved at a regular and unvarying rate. A pattern of genetic differences or distances among living species could be turned into a branching sequence in which branch lengths, it was assumed, were linearly proportional to time. The sequence could be calibrated using one fossil or geological event which could be dated. Sarich and Wilson used a blood protein called albumin and generated a hominoid tree which gave a date for a three-way split of chimpanzee, gorilla, and human of around 5 my, with the Asian orangutan roughly twice as different genetically, diverging at 10 my (Figure 4.2).

These dates contrasted very considerably with the notion of hominids diverging more than 14 my ago. Palaeontologists and morphologists rarely if ever see regular change in the anatomical systems they study. Many refused to believe the 'clock', and many are still reluctant to accept the idea that genetical patterns may be more informative about evolutionary relationships than morphology. The debate over the hominoid patterns which followed was vigorous and has occupied the last two decades, but it was only part of a wider set of disagreements about the nature of molecular evolution. How much was shaped by natural selection and how much was random or stochastic change? Although the argument is not yet settled, the shape of an answer has emerged.

DNA, the molecule carrying all inherited information, controls protein synthesis and through that cellular growth, development, differentiation, and integration. Selection acts on this portion of the DNA. But less than 10 per cent of all single-copy DNA (that present only once per cell, rather than several to many times) is 'expressed', that is, controls protein synthesis. The functions of most of the unexpressed portion are mysterious: some parts are known to vary

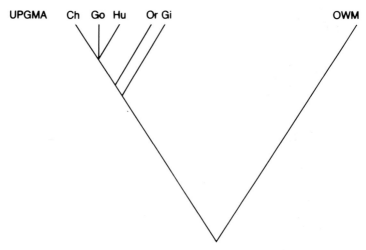

Albumin distances	Go	Hu	Ch	Or	Gi	OWM
Gorilla	–	4.5	8.0	10.5	11.0	32.0
Human		–	5.5	11.5	11.5	40.0
Chimp			–	9.5	13.5	37.5
Orang				–	11.0	40.0
Gibbon					–	38.5
O W monkey						–
Column average			6	10.5	11.8	37.6
Column range			3.5	2.0	2.5	8.0
% Ra/Av			58	19	21	21

UPGMA   Ch Go Hu   Or Gi   OWM

Fig.4.3. Matrix of albumin distances in hominoids and monkeys from Cronin (1975). The distances are quantitative measures of degrees of antigenic similarity between pairs of species. The branching sequence is derived from the matrix by the simple unweighted pair group method of analysis (UPGMA).

hardly at all across species and presumably have important functions. But probably the majority of unexpressed DNA is little subject to selection. Even at the level of proteins, not all changes have functional significance: some amino acids can change without affecting function. It now looks as though much change at the protein level and most at the DNA level is not greatly affected by natural selection, but 'ticks away' in the form of a steady stream of mutations, unseen by the editing of natural selection. We therefore do indeed have the makings of a 'clock', in fact of several clocks, some of which should work better than others.

Figure 4.3 shows a matrix of albumin distances. The numbers are a quantitative measure of the immunological reaction strength between the albumin of one species and antibody to albumin of another species. This is proportional to the amino-acid sequence differences of the two albumins, which is in turn a reflection of the sequence difference in the portion of DNA coding for albumin. Note that the distance or difference between the monkeys (OWM) and all hominoids is approximately equal, with an average value of

37.6 and a range of 8.0. This comparison, between a more distant 'outgroup', and a cluster of more closely related species of a monophyletic group, is called the rate test. The close similarity of these numbers, their uniformity, is difficult to explain unless there is some underlying regularity in the evolutionary process.

Another technique called DNA–DNA hybridization makes possible comparison of all single-copy DNA of two species. It has been hypothesized that the entire complement of single-copy DNA might evolve at a regular rate. Expressed portions (genes) would usually be subject to selection, directional or stabilizing at different times, and so might fluctuate in rate to a greater or lesser extent. Non-expressed DNA, which constitutes its greater fraction, will also have some parts which evolve slowly, others rapidly, some might vary in rate while others would not. But viewed overall, because single-copy DNA is so large (10^9 nucleotides or more in single-copy DNA, 10^3 in the average expressed portion of a gene) there is an averaging effect, producing a single overall rate which does not fluctuate: the uniform average rate hypothesis. First, the species under consideration must hybridize approximately equally: this eliminates situations where in one species a significant portion of DNA might have evolved very rapidly. If per cent hybridization values are comparable for the species under consideration, the rate test can then be used. In the case of hominoids, both per cent hybridization and the rate test applied to per cent sequence difference show remarkably unifrom patterns. Figure 4.4 gives new hominoid data generously supplied by Charles Sibley and Jon Ahlquist. Clearly, DNA in this group is evolving at a regular rate relative to time, although there is no way of knowing whether or not this is linear. Until we know more about the behaviour of the vast majority of DNA we shall not be able to propose a realistic model for DNA evolution.

Nonetheless, these hominoid data clearly show some monotonically regular pattern of change, and, equally clearly, difference can be equated to genealogical distance. The fact that humans and chimpanzees are most similar therefore means that they are most closely related. Although the DNA differences are not obviously linear relative to time, I assume that relatively small differences are relatively linear and that the branch lengths in the evolutionary tree are proportional (enough) to time. These patterns are the best we can hope for; they are similar to those generated earlier from the albumins (Figure 4.3), although branch lengths differ a little, and the human–chimpanzee–gorilla trichotomy is resolved. The use of these DNA data irritates both many morphologists and palaeontologists (it is a phenetic and not a cladistic technique, and claims extremely 'clocklike' behaviour), and many biochemists (it is not a biochemically elegant technique). We shall, however, proceed with them, and see how we can fit fossil data to the branching patterns.

We have many more hominoid fossils than even five years ago, enough to provide us with the beginnings of a realistic framework for hominoid evolution. Fossils can now be more accurately dated, and, more important, body parts other than jaws and teeth – ambiguous indicators of evolutionary relationships – have been recovered. In particular, parts of the face are proving

DNA – DNA distances	Ch	Hu	Go	Or	Gi	Ba
Gorilla	–	1.6	2.2	3.6	4.8	7.4
Human		–	2.3	3.6	4.8	7.3
Chimp			–	3.5	4.8	7.1
Orang				–	4.8	7.4
Gibbon					–	7.1
O W monkey						–
Column average		1.6	2.2	3.6	4.8	7.3
Column range			0.1	0.1	0	0.3
% Ra/Av			4	3	0	4

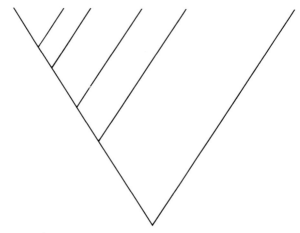

Fig.4.4. Matrix of DNA hybridization distances in hominoids and monkeys (Sibley, pers. comm.). The distances are equivalent to per cent sequence difference between species pairs. Note how much more uniform the data are than from albumins (Figure 4.3). The branching sequence is derived from the matrix by the simple unweighted pair group method of analysis (UPGMA).

very interesting in unravelling relationships. Figure 4.5 shows patterns in the hominoid face when viewed in midsection. Shape and disposition of the palate, of the incisive canal which runs through the palate in the midline and contains nerves and blood vessels, and of the premaxilla or lower face have been examined in detail in monkeys and hominoids by Steven Ward. Gibbons and monkeys show one pattern (*a*), gorillas and chimpanzees a second (*b*), orangutans a third (*c*), and humans a fourth (*d*). The hominid fossil record is sufficiently dense to show that the human pattern evolved from an early *Australopithecus* pattern identical to that of the chimpanzee. Figure 4.6 shows the likeliest evolutionary sequence for these facial patterns: (*a*) to (*b*) to (*c*), and this hypothesis can be used, in conjunction with other anatomical systems, to help sort Miocene hominoids (the Miocene is the time between 24 and 5 my ago) into correct species, and to arrange those species on the tree whose framework is derived from DNA hybridization analyses.

Figure 4.7 shows the distribution of the better-sampled fossil apes. *Proconsul* from Kenya ranges between 22 my and 18 my and was a generalized and

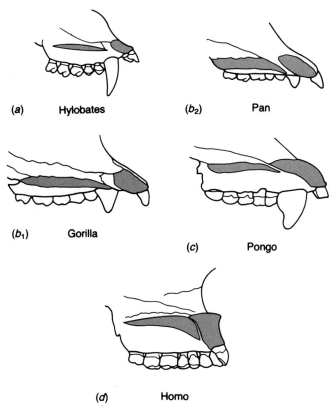

(*a*)   Hylobates   (*b₂*)   Pan

(*b₁*)   Gorilla

(*c*)   Pongo

(*d*)   Homo

Fig.4.5. Mid-line cross-sections of lower face in living hominoids
showing premaxillary–palatal morphology.

mainly quadrupedal animal, mostly living on fruits in forests or woodlands, predominantly in the trees but probably coming to the ground occasionally. *Proconsul* is anatomically primitive – for example in the face – and is close to the ancestry of all living hominoids. Several fossil samples from the 14-my to 18-my period suggest that the split between lesser apes (gibbons) and larger hominoids (chimpanzee, gorilla, human, orangutan) occurred during this time. New fossils of *Kenyapithecus* and a new and as yet unnamed genus from Buluk in Kenya differ from *Proconsul* in ways suggesting relatedness specific-ally to large hominoids. For example, they are more evolved in facial and tooth anatomy. By 12 my ago, large hominoids had spread out of Africa. The later Miocene, from 12 my to 7 my ago, has produced many new hominoid fossils, and it now looks as though there are at least four distinct forms present, as different as many living genera. *Rudapithecus*, possibly a synonym of the poorly known *Dryopithecus*, retains the primitive facial features found in *Proconsul*. *Ouranopithecus* from Greece is facially like the living African apes, as is *Kenyapithecus*, but the two fossils differ in other ways. The hominoids from China have been called *Sivapithecus* and *Ramapithecus*, but they represent one species which is different overall from all others and which deserves its own new name. This brings us to *Sivapithecus*.

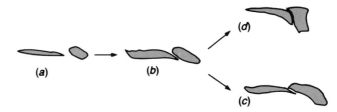

Fig.4.6. Hypothesized evolutionary sequence in hominoid pre-maxillary–palatal morphology (after S. Ward).

Once used to cover a wide range of material, the names *Sivapithecus* and *Ramapithecus* are now best restricted to specimens only from Turkey, Pakistan, India, and Nepal dating between 12 my and 7 my. New material makes it highly probable that only one species is actually represented at any one time, and that '*Ramapithecus*' had small canines because they were female *Sivapithecus*. In facial anatomy *Sivapithecus* (Figure 4.8) differs from all other Miocene hominoids and closely resembles *Pongo*, the living Asian orangutan (Figure 4.9). *Sivapithecus*, including *Ramapithecus*, indeed shares a number of tooth features with *Australopithecus*, but these are probably either retentions from the ancient hominoid conditions and/or parallelisms. The facial resemblances to *Pongo* are detailed, and there are plausible reasons for believing that these features are more evolved, allowing us specifically to link *Sivapithecus* and *Pongo*. In the same facial parts early *Australopithecus* resembles the chimpanzee *Pan* (Figure 4.10).

This linking of *Pongo* and *Sivapithecus* gives us a good calibration point for our hominoid tree, making the divergence of the orangutan lineage at least 12 my

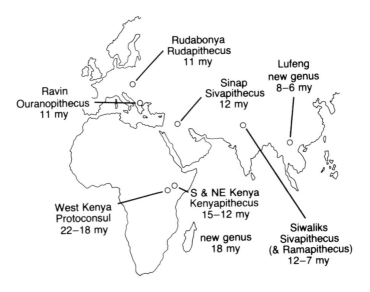

Fig.4.7. Distribution of better Miocene hominoid material.

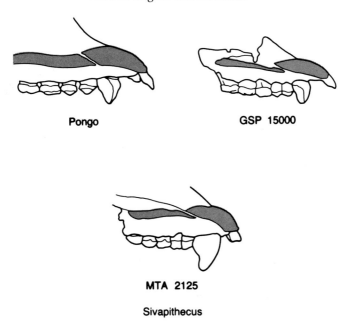

Pongo

GSP 15000

MTA 2125

Sivapithecus

Fig.4.8. Mid-line sections of lower face in orang (*Pongo*) and *Sivapithecus* showing similarity in premaxillary–palatal morphology (after S. Ward).

ago, the age of the oldest recognizable *Sivapithecus* (Figure 4.11). If all these arguments are correct, the divergence age of gorillas from the 'African' hominoid branch can be set at between 7 and 8 my and the chimpanzee-hominoid split at 5–6 my.

What were these Miocene hominoids like as animals? *Proconsul* and to a lesser extent *Sivapithecus* are now known from a sufficient range of body parts to give us some useful clues. They were dentally very size-dimorphic. That is, males and females differed very markedly in the size of the cheek teeth, more so than in living primates, and in body-size dimorphism they were also at or beyond the 2–1 limits seen between the extreme living male and female primate body sizes. Judging from tooth wear they ate mostly fruit. Limb bones show that *Sivapithecus* was apelike overall: long-armed, and climbing, hanging, and swinging using the arms, with the body frequently vertical. Legs were probably also long, and used in vertical climbing and bipedal walking. Feet were capable of powerful grasping. At home in the trees and no doubt occasionally on the ground, *Sivapithecus* was apelike, but in its total behavioural and movement repertoire unlike any living hominoid.

Figure 4.12 shows *Sivapithecus* as generalized and different from the more specialized African ape knuckle-walkers, orangutan contortionists, and hominid bipeds. *Sivapithecus*, although not the common ancestor of all the large hominoids, was probably not too different from that ancestor. Unfortunately, the African fossil record is not yet good enough to give us the necessary information, and the critical 5–9-my period during which hominoids, chimps,

[101]

and gorillas diversified is essentially a blank. We do not know whether pre-hominids would have most closely resembled chimpanzees, *Australopithecus* or a more generalized *Sivapithecus* type.

Fig.4.9. Partial face of *Sivapethicus*, an 8-my ape related to the orangutan, Petwar Plateau of Pakistan (courtesy of W. Sacco and D. Pilbeam).

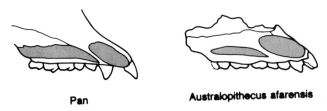

Pan   Australopithecus afarensis

Fig.4.10. Mid-line sections of lower face in chimpanzee (*Pan*) and *Australopithecus afarensis* showing similarity in premaxillary–palatal morphology (after S. Ward).

The origin of hominids is one of three critical steps in hominid evolution. To describe it adequately we would ideally need to know both 'before' and 'after' stages. 'Before' is unknown, while 'after', though generally considered to be early *Australopithecus*, is also actually effectively unknown, or it is if the chimpanzee–hominid divergence occurred over 5 my ago; *Australopithecus* is not well known until less than 4 my and judging from subsequent hominid evolution there is 'room' for at least one earlier and more primitive species.

Despite the absence of fossil data there has been a great deal of speculation about hominid origins. There is agreement now that the critical step involved locomotion, and saw an increase in the frequency of bipedalism in the

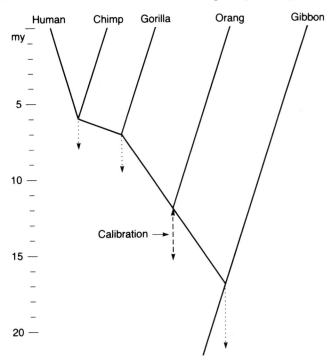

Fig.4.11. Hominoid branching sequence based on DNA hybridization data (Figure 4.4), calibrated assuming *Sivapithecus* is ancestral to *Pongo* and oldest *Sivapithecus* is 12 my old.

[103]

David Pilbeam

Fig.4.12. Hominoid tree showing diversity of living large hominoids. Their common ancestor is depicted as a generalized arboreal ape similar to *Sivapethicus*.

repertoire of movements. There has been much speculation why. Because of tool-use either as digging sticks or weapons? Because of food transport, either meat or potatoes? Because mates carried home the bacon to their bonded mates? There is a growing consensus that the achievement of significant amounts of bipedalism involved food and feeding behaviour and was probably related to feeding on the ground. Whether or not it initially involved transporting food or other objects, or whether this was a consequence of becoming more bipedal, is unclear. We need fossils, and many of them.

First hominids: Australopithecus and early Homo

Human and chimpanzee lineages had split at least 5 my ago, probably in Africa, because we are sure that hominids were confined there for over 4 my to around 1 my ago. The oldest well known hominid is *Australopithecus afarensis*,

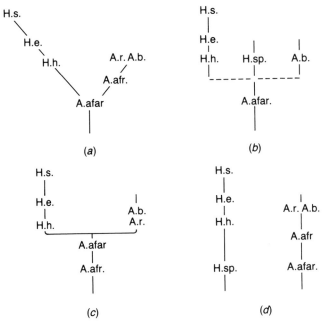

Fig.4.13. Alternative evolutionary trees for hominids: A afar, *Austra-lopithecus afarensis*; A. afr., *A. africanus*; A.r., *A. robustus*; A.b., *A. boisei*; H.h., *Homo habilis*; H.sp., *Homo* species; H.e., *H. erectus*; H.s., *H. sapiens*.

dated between 4 my and 3 my, from east Africa. This is seen by many as the ancestral hominid, with one lineage of descendants (see Figure 4.13) leading through South African *A. africanus* (3–2 my) to the extinct specialized South African *A. robustus* (2–1.5 my) and the hyper-robust east African *A. boisei* (2.5–1.5 my). Another runs through east African *Homo habilis* (2 my–1.6 my) to *H. erectus* (1.6–0.3 my) to *H. sapiens*. In the latter lineage, brain size increases and stone tool-making appears and becomes more complex. There are alternative trees based on the same data (Figure 4.13), for example, showing *A. africanus* on the line leading to *Homo* rather than to robust australopithecines. Some believe that *A. afarensis* itself represents two species rather than one, and others that *H. habilis* also consists of two species.

These differences of opinion about basic taxonomy (how many species) and evolutionary relationships are important. But they are inherently difficult to resolve, particularly the issue of precise relationships. Relationships are ambiguous because the species under consideration are so similar. Yet this in turn means that biological differences between them are perhaps not so great, so precise relationships will not make much difference to the basics of the story. *A. afarensis* and *A. africanus* are very similar to each other except in small features of teeth and jaws, as similar in most features as two species of baboon, or the two chimpanzee species. There are two kinds of baboons, common baboons, *Papio cynocephalus*, and hamadryas baboons, *Papio hamadryas*. Their teeth and bones are almost indistinguishable but their behaviour is very different. Common baboons live in promiscuous multi-male social groups while hamadryas baboons live in polygamous groups dominated by a single breeding male: very different social and reproductive behaviour, but unpredictable from anatomy. The two species of chimpanzee, the common chimpanzee *Pan troglodytes* and the bonobo *Pan paniscus*, differ slightly in morphology, probably more than the two australopithecines, and their behaviour differs, too, but less than between the two *Papio* species.

We cannot therefore expect too much detail from the fossil record. Statements are usually likely to be at the level of the kind of generalizations one can make about living genera rather than species. Nor should we argue too much about matters which cannot be settled or which do not much affect the principal knowable issues of behavioural evolution.

Until about ten years ago the general view of the australopithecines – at least the smaller ones – was that despite small brains they were behaviourally rather like us: open country hunter-gatherers, using tools, utilizing caves as home bases, perhaps with a human-like family organization. That view has changed, both for good empirical reasons – the data cited as supporting carnivorous behaviour do not – and also because the perspective pendulum has moved. It has shifted to a view of past hominid species as more unlike us than like us, or at least to a position that takes that as the more sensible *a priori* expectation. But behavioural reconstruction of the australopithecines is more realistic now than it was, partly thanks to better fossils, partly to better studies of living animals and ecology, especially the apes.

The last two decades have seen another biological revolution in addition to the molecular one, this in the understanding of behavioural ecology and its principles: the 'sociobiological revolution'. Although this has been contentious at times, most behavioural biologists now tend to look at the behaviour of individuals from the perspective of what maximizes Darwinian fitness, and what promotes survival and reproduction. Which behavioural strategies do individuals pursue to maximize the likelihood of gaining food and mates, and minimize the likelihood of being eaten? How can the common principles underlying animal behaviour be used to explain the diversity of animal behaviour? Particularly important have been field studies of the four barely surviving large apes, which have produced a wealth of important information about diverse social behaviours and their links to ecology and to anatomy. Sense is now being made of this rich diversity. We are beginning to look more to ape behaviour than to that of contemporary humans for clues to deciphering the australopithecine puzzle. But rather than modelling australopithecine behaviour analogically (transferring all or part of extant ape behaviours to the past), we do so strategically. We want to reconstruct the critical determinants of past social organization, such as food and its distribution (although food is not the only one), and then infer possible behaviour patterns in the australopithecines based on our knowledge of their anatomy and palaeoecology, and our understanding of living systems.

The general consensus now is that the australopithecines were the hunted rather than the hunters and were basically apelike in their individual and social behaviour. Anatomically they are quite well known, especially the smaller and earlier species, *A. afarensis* and *A. africanus*. We know from associated plant and animal fossils that they were part of open habitat communities: savanna and woodland rather than forest. Clearly, they were effective bipeds, and moving on the ground must have been important. Limb proportions – they had relatively longer arms and perhaps shorter legs than *Homo* – and some anatomical features of arms and hands imply that they still climbed trees, perhaps for food or protection or to rest. Early hominids would also have trekked bipedally on the ground between food sources, perhaps because being upright made this energetically efficient. Chimpanzees spend much time on the ground, especially in open habitats, even though they feed mostly in trees. But chimpanzee ground travel, knuckle-walking quadrupedalism, is energetic-ally more costly (by about 50 per cent) than either 'regular' quadrupedalism (of baboons, for example) or human bipedalism. This is because chimpanzees need to retain the anatomy of skilled tree climbers: short and stiff lower backs, short legs, long feet with long toes and a divergent grasping big toe. None of these features allow efficient bipedalism. Given that hominids evolved from apelike ancestors, their most efficient 'option' for travelling on the ground would be bipedalism, achieved by increasing the amount of bipedalism already present in the repertoire of ancestral apes.

Male–female body-size difference was perhaps as much as 2:1 in species of *Australopithecus*, and in this they resemble most extinct and living large hominoids except the chimpanzee. In *Pan* males are only about one-third larger than females, rather than twice the size. It has been suggested that chimp males are not bigger because they have to cover large areas, to monitor females and food and to defend range boundaries, and knuckle-walking quadrupedal-ism is relatively inefficient for long-distance trekking.

Australopithecus canines are smaller and less dimorphic than those of apes but larger than later *Homo*. Cheek teeth are larger relative to body size than in apes, with very thick enamel caps. Surprisingly, microscopic tooth surface wear – which reflects the kinds of foods being chewed – is basically like that in chimpanzees, suggesting a mostly plant diet made up mainly of fruits and occasional harder seeds or nuts. But judging from palaeoecology and anatomy, this was probably not the ripe forest fruits favoured by chimps. Ape and australopithecine foods would have been somewhat different; precisely how different remains an important question.

If we could watch the early australopithecines from our time machine, what might we see? We have a range of suggestions: Raymond Dart's hunters; Owen Lovejoy's idea that bipedalism and monogamy evolved together, bipedal males carrying food to provide for their bonded and less mobile mates; Adrienne Zihlman's expectation, inspired by the genetic closeness of humans and chimpanzees, that australopithecine social organization would have been flexible, chimp-like, but focused around females. It is easier to say what past species did not do than what they did, because the interesting past species are *un*like living ones. This is why we have to pursue their understanding from a

[107]

reductionist perspective based on our knowledge of living systems. But such understandings are rarely unambiguous. We will have to settle for a 'pointilliste' approach: best at a distance and without close detail.

Australopithecines probably foraged mostly as vegetarians in small social groups of several females and their offspring tolerating and focused around an adult male. Small brains imply apelike maturation rates, confirmed by tooth-development patterns. They were probably not monogamous because monogamous primates show little sexual dimorphism, and 'marriage' might have been beyond their symbolic capabilities. They were plant eaters, but it is unclear what they ate, although it is likely to have been often dispersed. Although they apparently did not flake stone, they probably used tools as much as chimps do or more: stone, wood, vegetation would have been utilized for both extractive foraging and aggressive display. Australopithecine hands seem better adapted for tool-use than chimp hands. Their communications systems probably did not involve speech. It is unfortunate that we do not understand more about chimpanzee communication in the wild: it is clearly complex, and presumably australopithecines would have been equally or more so.

One of the most interesting questions is why the australopithecines were bipeds, and knowing the answer will have at least some impact on our thinking about the origin of bipedalism in early hominids. (But remember that the function of a character often changes through time: for example, even if we could be sure that australopithecines carried objects, that would not mean that bipedalism necessarily evolved as an adaptation for carrying.) Perhaps they were bipedal because they were less committed to life in the trees than chimpanzees. They could then evolve (or retain) long legs and flexible lower backs which would be impossible for a large tree climber. Australopithecine bipedalism may or may not have been less efficient than human walking, but it was certainly more efficient that any kind of chimpanzee ground-walking. So it could be an adaptation for covering long distances in a search of food and other resources (mates) and to defend territory. It is unclear whether carrying was part of the original adaptation for bipedalism, or was facilitated by it. Too close and we lose detail.

Some of the later australopithecines increased in body size (*A. boisei, A. robustus*). Tooth size and brain size also increased but remained approximately proportional to body size. These species lived in east and southern Africa in more open habitats than the earlier australopithecines, and size increase might perhaps reflect a generally lower quality diet. They are extremely interesting creatures in their own right, becoming extinct over 1 my ago. They tend to be overshadowed because of greater interest in the other descendants of early *Australopithecus*, the first *Homo* species.

By 2 my ago (and when the record is better sampled it will probably be closer to 2.5 my) new and distinctly different hominids had evolved. One was a large-brained species, *Homo habilis*. The other is less well known, and believed by some to be females of *habilis*, but the consensus is that another species of *Homo* was also present around 2 my ago. At this time and earlier the first well-dated archaeological traces appear in the form of chipped stonecores, sharp flakes, and broken bone. Where and when this evolutionary change occurred we do not known, nor how quickly, although it was very probably in Africa between 2.5 and 2.0 my ago. The ancestor of *habilis* is unknown or unclear and more collecting is needed in east Africa prior to 2 my.

Until recently *H. habilis* was conceived as behaviourally rather like us – or at least a certain vision of us – hunting, sharing food, foraging from and returning to home bases, perhaps living in family groups. That view is shifting and *habilis* is becoming more enigmatic and harder to reconstruct. *Homo habilis* and smaller *Australopithecus*, for example *A. africanus*, as shown in Figure 4.14(*a*), and (*b*), had faces and teeth similar in size, suggesting that their diets were probably similar and composed of predominantly plant food. But they differ in brain size, perhaps in external brain anatomy (if the inside of the skull is any indication) and perhaps in internal organization (if external brain anatomy is any indication), *H. habilis* having a significantly larger brain (600–800 cm^3) than *A. africanus* (400–500 cm^3). The little that is known of the limbs and trunk in early *Homo* show basic similarities to humans, though they are more robust, and some differences from australopithecines. A ground-dwelling bipedal pattern essentially identical to ours is indicated, well adapted to long distance, endurance trekking and running. There are disagreements about the precise meaning of the archaeological traces, but at least they imply more meat-eating than in australopithecines: *habilis* was edging out of the herbivore niche, perhaps competing with other scavengers for carcasses left behind by carnivore hunters, perhaps hunting small game as well. The extent of food sharing is unclear, and probably considerably less than in modern humans.

Note that brain expansion comes at least 2 or 3 my after the evolution of a significant degree of bipedalism, so being upright did not cause brain expansion – in the sense of being both necessary and sufficient – as palaeoanthropologists have sometimes argued. With the recognition that the australopithecines and early *Homo* were both different from us and different from each other, we begin to understand that human features did not evolve precisely together and in step, but as a mosaic instead: there were many stages to becoming human.

Evolution in the genus Homo

Homo habilis was replaced over 1.5 my ago by a slightly larger brained hominid, *Homo erectus* (Figure 4.14(*c*)). Where, when, and how rapidly this happened are again unknown, although Africa is the likeliest place and rapidly the likeliest tempo. *H. erectus* had a brain over half the size of ours (800–1000 cm^3 versus 1100–1600 cm^3) and almost double that of the australopithecines. Why the brain expanded is unclear, and a question that will perhaps never be satisfactorily answered; but knowing much more about brain structure, behaviour, and communication in living primates, including humans, will be helpful. The consensus is that these hominids had linguistic and symbol-making abilities clearly greater than those of apes and australopithecines. One

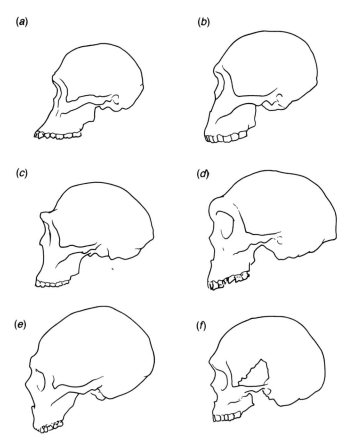

Fig.4.14. Hominids. a, *Australopithecus africanus*; b, *Homo habilis*; c, *H. erectus*; d, archaic *H. sapiens*; e, neanderthal, f, modern *H. sapiens*.

of several consequences of having a larger brain is slower maturation: larger-brained animals take longer to grow up, and they reproduce more slowly. There was probably a marked shift in life history patterns and feeding behaviour in these archaic *Homo* species, away from more apelike patterns. But more about life history patterns later.

Anatomical evidence from face, jaws, and teeth of *H. erectus*, which are smaller than those of *H. habilis* (Figure 4.13(b)), as well as archaeological evidence, suggests a change to a diet with somewhat more meat. Relevant to this is a fascinating piece of fossilized behaviour, much of a *H. erectus* skeleton found by Richard Leakey's group in East Turkana in Kenya, and over 1.5 my old. This fossil preserved much of its original histology in microscopic detail. A cross-section through the femur shows both normal bone in the very thick cortex (thinner in a modern human) and disorganized pathological bone forming a crust around the entire outer surface. All the long bones were similarly affected. Death would have followed the processes responsible for generating this pathological bone, and it has been suggested that this was caused by a massive overdose of Vitamin A, probably ingested in a single meal. It has also been suggested that this meal included raw carnivore liver: carnivores concentrate Vitamin A in their livers while plant eaters do not. Perhaps we see in this specimen not only evidence for meat-eating but also for interference competition between hominids and carnivores. The extent to which meat in *H. erectus* diets came from scavenging or hunting is hotly debated.

Another splendid *H. erectus* specimen out of Africa, a young male skeleton from west of Lake Turkana found in 1984, is the most complete individual of this species. Body proportions and stature were human, but the hips were very narrow. Although the specimen was male, we can predict with some confidence that females would also have had relatively narrow hips and therefore small birth canals. This implies the birth of small and immature infants after a relatively short gestation, dependent on their mothers and reducing their mothers' mobility. It also implies other effects on the behaviour and interactions of males as well as females.

Homo erectus was the first hominid species to achieve a wide distribution outside Africa, as far as eastern Asia, the exodus from Africa probably occurring around 1 my ago. It was also probably the first user of controlled fire, although this is debated, and exactly when it happened is unclear. *H. erectus* made tools in a more regular and patterned way than the first tool makers. Their finely made stone implements imply manipulative and mental skills that are more human than earlier hominids. *H. erectus* and its industries lasted from over 1.5 my ago to less than 0.5 my ago without much change, implying a degree of behavioural stability which was surprising when it first became apparent to palaeoanthropologists.

Around 0.3 my ago, hominids with somewhat larger brains (1100–1300 cm^3) and slightly differently shaped skulls began to appear: so-called archaic *Homo sapiens* (Figure 14.4(*d*)). Overall they were still very similar to *H. erectus*. To distinguish them from us, modern *Homo sapiens*, the 'archaic' modifier is added. The best-known archaic *sapiens* came from Europe and west Asia: the neanderthals (Figure 4.14(*e*)). Archaic *sapiens* produced more diverse and better-made tools than *erectus*, using smaller flakes rather than larger cores, and they were probably more effective hunters: many of their later tools were made to be hafted as spears and knives.

However, despite their sapient label, despite attempts to 'launder' them, I do not think they should be called *Homo sapiens* because morphologically and, by inference, behaviourally they differ markedly from modern humans. Although they were more like us than like the australopithecines, archaic *Homo* (*H. habilis* and *erectus* plus archaic *H. sapiens*) generally had more massive skeletons and heavily muscled bodies, implying high activity levels and other significant biological differences from us. It has been suggested that this might be part of a biological pattern involving persistent running, allowing hominids without great intrinsic speed and strength or elaborate technology – but good endurance – to become predators of small game: running them down and wearing them out. Some archaeologists believe that tools and bones left by

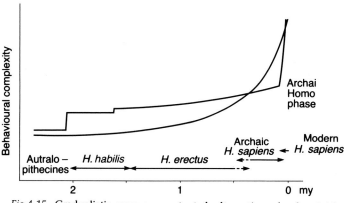

Fig.4.15. Gradualistic versus punctuated alternatives for hominid behavioural evolution.

archaic *Homo* imply they were scavengers and hunters, at most, of small game, incapable of planned logistic big-game hunting in the way modern humans are. Further, it is argued that food preparation and sharing was less elaborate, and that human-type home bases – foci for many different activities – were not yet developed.

The pelvis in neanderthals (not known in other archaic *sapiens*) was wide in both sexes, implying the birth of a very large infant. There is disagreement over whether this is merely because neanderthals were themselves very stocky and powerfully built, or because infants were born larger and more mature after a gestation of 11 or 12 months. On this issue, as with others, the jury is still out, although there is the distinct possibility that, even as recently as 100 ty ago, our ancestors were still in important ways distinctly non-modern.

This is a good point at which to mention one of the most interesting debates in palaeoanthropology: the trajectory of behavioural education in *Homo* (Figure 4.15). Was it gradual, progressive, and accelerating – which is still probably the 'traditional' view? Or was it 'stepped', with a long-lasting pre-*sapiens* phase that changed little and was fundamentally non-modern, even late in the sequence? How 'modern' in behaviour were the archaics? The density of human fossils and archaeological data is better for late archaics like neanderthals than for any other phase in human prehistory. Let us hope that these questions can be answered, because if they cannot then reconstructing earlier stages will also be impossible.

Both anatomy and archaeology surely show contrasts between archaic and modern humans. I do not intend to paint neanderthals as dumb apes: apes are not dumb, and neither is that the only choice, because there is a great deal of 'behavioural space' between apelike and humanlike systems that was filled by our ancestors. But everything points to a system in which communication, presumably speech, was less effective than in modern humans, and the amount of stored and transmitted information less. Archaic humans would have been less symbol-competent and more biology-bound than us.

A substantial evolutionary change comes between archaic and modern *Homo*

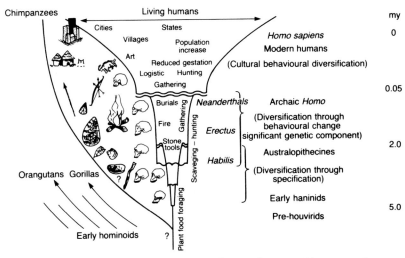

Fig.4.16. Diagrammatic summary of major features of human evolution (courtesy of Jeanne Sept).

sapiens (Figure 14.13(*f*)). It occurred over the period between 130 ty and 30 ty
ago, ending with the appearance of humans like us, with less massive
skeletons, less muscular bodies, rounder heads and flatter faces. Judging from
their archaeological traces, their behavioural potential was also like ours by at
least 35 ty ago. The cause of this last major change in human evolution is
obscure, and there is considerable disagreement about its tempo and mode:
when and where it happened and what microevolutionary pattern was
followed. Opinion ranges from the view that the archaic to modern transition
occurred essentially throughout the occupied Old World, to the claim that
modern humans appeared first in one geographically restricted part of the
archaic human range, probably Africa, and spread from there. Genetic studies
of living populations suggest that Africa may indeed have been the source of
modern humans, and the fossil record can be interpreted to support this view.

There is general agreement that language played a critical role in this
transition; what is seen might be the result of the final evolution of fully
modern language capabilities. With these modern humans we get the first
evidence for recognizably modern behavioural patterns of many kinds – from
cave painting and sculpture, to sophisticated tools and elaborately planned co-
operative hunting, to increased population size and density and broadened
geographical and ecological range. It is here, finally, that all the truly 'human'
characters come together for the first time.

Conclusions

We have seen great progress in the last two decades in putting together a
reasonable sequence, and this is summarized in Figure 4.16. We still need a

better record of the critical stages; some of them are still not documented at all. We need to improve behavioural–ecological reconstructions. The study of the past is the study of a kind of present; the better we understand the present, the clearer our insights into the past. This is why one area in which palaeoanthropology needs to advance is in understanding living species and the way their behaviour is organized. We must also learn to tell the story from the beginning to the end, rather than the other way round from the selective bias of the present. Each evolutionary stage was indeed a platform on which the next developed, but each stage existed and must be seen on its own terms, not merely as the source of what followed. Australopithecines did not know that *Homo habilis* was the next step, nor did neanderthals feel us breathing down their necks. There was no single point at which we became human. Many 'human' qualities seem to have evolved surprisingly late, which is one of the most interesting consequences of recent work. This has implications for theories of human evolution, particularly those depending on early hominids being *like* us in critical features.

Paraphrasing George Orwell, all species are unique, but humans are more unique than other species. We are all aware of the problems of explaining unique phenomena; we are stuck with unique causes. Becoming human was an accretionary process of many stages involving odd events – language, bipedalism, meat-eating, fire-use, shorter pregnancies; all were probably very unlikely events and certainly not inevitable. Slowly we are improving our understanding, but some of those interesting events will remain forever unexplained. This bothers some people. But I am reminded of what the great physicist Richard Feynmann said: 'I can live with doubt and uncertainty. I think it's much more interesting to live not knowing than to have answers which might be wrong.'

Further reading

Cartmill, M., Pilbeam, D., and Isaac, G. *One hundred years of paleoanthropology*, Amer. Sci., 1986.
Delson, E. (ed.) *Ancestors*. Alan Liss, New York, 1985.
Gowlett, J. *Ascent to Civilization*. Knopf, New York, 1984.
Jolly, A. *The Evolution of Primate Behaviour, 2nd edition*. Macmillan, New York, 1985.
Lewin, R. *Human Evolution*. Freeman, New York, 1984.

[5]

Origin of social behaviour

John Maynard Smith

Introduction

The first part of this chapter is concerned with the origins of social behaviour among animals. The theoretical difficulty is obvious. Darwinian natural selection will favour traits that ensure individual survival and reproduction. How, then can we account for co-operative behaviour, in which one individual helps another? The difficulty is particularly acute when one animal helps another at the expense of its own chances of reproduction. Yet such self-sacrificing behaviour certainly occurs, most notably among the social insects. I have nothing particularly new to say about this topic: instead, I will summarize the picture that has emerged during the past 20 years, mainly from the work of others.

I then turn to the distinction between the mechanisms of social behaviour in man and other animals. I do this by discussing a particular 'game', the social contract game, which I think captures the essential differences. These are the existence of speech, and of a particular kind of self-consciousness. However, although the social contract game does bring out the differences between man and other animals, it is not a particularly satisfactory model of human society, mainly because it ignores the differences between people in terms of the actions open to them. I therefore conclude the essay by discussing the significance of these differences. This leads to the idea that a crucial problem in understanding human societies concerns the ways in which individuals associate together in groups.

Social behaviour in animals

In analysing the selection pressures that can lead to the evolution of co-operation among animals, a good starting-point is the 'trait group' model (Figure 5.1) proposed by D.S. Wilson (an identical model was proposed independently by Matessi and Jayakar) and by Cohen and Eshel. The members of a population are supposed to mate randomly. The offspring then associate in

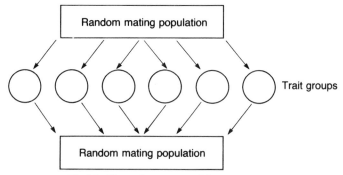

Fig.5.1. The trait group model.

'trait groups': it is while they are in these groups that natural selection acts upon them. After selection has acted, the groups break up, and the surviving individuals come together to mate and produce the next generation. In the simplest version of the model, individuals are of two kinds, co-operators (C) and defectors (D). Each C individual performs some act which has the effect of reducing its own fitness by c ('cost'), and of increasing the fitness of all other members of the group by $b/(n-1)$, where b stands for 'benefit', and $n-1$ is the number of other individuals in the group, between whom the benefit must be shared. 'Fitness' means probabiliy of surviving. The point about this way of calculating fitness is that costs and benefits combine additively: the significance of this will emerge later.

   If we ask whether co-operation will increase in a large population, consisting of many groups, a very simple answer emerges. Provided that:

(i) individuals associate into groups randomly, and
(ii) costs and benefits combine additively

then it can easily be shown that the only thing that matters is the cost, c. If there is a positive cost, co-operation will decrease in frequency: the value of b is irrelevant. The reason why this is so is as follows. When a co-operator acts, it reduces its own fitness by c, and increases the fitness of a random sample of the population. The second of these two effects has no influence either in increasing or decreasing the frequency of C, so only the value of c matters. At first sight, this is a rather boring conclusion. Co-operative acts, in the sense of acts that help others, can spread only if the same acts increase the fitness of the performer (c negative). However, a more interesting conclusion is as follows. If we want to change the model so that 'altruistic' acts (i.e. acts that cost something to perform) can evolve, we must change one of the two assumptions listed above. Either association into groups must be non-random, or costs and benefits must combine non-additively. I will consider these alternatives in turn.

Non-random association into groups

If, when a co-operator acts, the benefits of the action were received preferentially, or exclusively, by other co-operators, then the argument above fails, and the value of b becomes relevant. The commonest reason why association into groups is non-random (or, more generally, why interacting neighbours are similar: the existence of distinct groups is unnecessary) is that groups consist of genetic relatives. The modern approach to such cases can be traced back to Hamilton's (1964) papers on 'The genetical evolution of social behaviour'; in explaining it, I shall follow the approach used by Dawkins (1976). The central point is that individual organisms are mortal, but their genes (or, more precisely, the information coded in their genes) are potentially immortal. Our concern, therefore, is why genes causing co-operative behaviour should increase in frequency.

Consider first the straightforward case of parental care. Since bodies are mortal, a gene will increase in frequency only if individuals in which it finds itself leave, on average, more children. If parental care causes an individual to leave more surviving offspring, then a gene causing parental care will increase in frequency, essentially because those offspring carry copies of the gene. But why should this argument apply only to actions that increase the number of offspring, or of direct descendents? After all, it is not the actual physical gene that is transmitted, as a man might leave his watch to his son: it is the information coded for by the gene. That information is also present in the genes of collateral relatives, such as brothers and cousins, inherited from a common parent or grandparent. We can therefore expect the spread of genes causing actions that increase the fitness (i.e. expected number of offspring) of relatives, whether they are direct descendents or not. Hamilton gave the condition for the spread of such genes in his inequality, $rb > c$, where c is the cost to the actor, b the benefit to a relative, and r a measure of their genetic relationship: b and c are measured in expected number of children.

The importance of this process for the evolution of social behaviour in animals is shown by the fact that all complex animal societies are composed of relatives. In some cases, it has been shown that degree of co-operation does depend on degree of relationship, and that Hamilton's inequality is satisfied. Trivers gives a recent review. Here I will comment briefly on two objections that have been made to the analysis of animal societies in terms of genetic specific behaviour. I think that this is a misunderstanding of what is required by the theory. Of course, a single gene does not determine a complex behaviour, regardless of the environment or the rest of the genotype. But that is not necessary. What is needed is that the substitution of one allele by another should make an animal somewhat more likely to perform some act, given the circumstances typical for that species. Empirically, what must be shown is that members of a species differ in behaviour, and that the differences are, at least in part, genetically caused. There is plenty of evidence of this kind; I will give one example. Two populations of the spider, *Agelenopsis aperta*, living in a woodland and in desert grassland, respectively, differ behaviourally in a number of ways, including the size of territory they will defend, the persistence with which they will fight for their webs, and their willingness to bite an opponent. When members of the two populations were raised from the egg stage in the laboratory, the differences persisted. By crossing them, and

John Maynard Smith

Figure 5.2. *Additive and synergistic fitness interactions.*

Type of pair

	C, C	C, D	D, D
Additive	4, 4	1, 5	2, 2
Synergistic	8, 8	1, 5	2, 2

Pay-off matrices

(a) Additive

	C	D
C	4	1
D	5	2

D is only ESS
'Prisoner's dilemma'

(b) Synergistic

	C	D
C	8	1
D	5	2

C and D are
both ESS's

Fig.5.2. Additive and synergistic fitness interactions.

raising a second generation of hybrids, it has been found that each of the differences is caused by genes at several loci, and not at one.

A second objection, also based on a misunderstanding, is that animals cannot be expected to calculate genetic relationship. This is in a sense true, but it is irrelevant. What is required is that an animal should behave in a way that is appropriate for the degree of relationship that typically holds. For example, two nestlings in the same nest are typically full sibs or half-sibs (adultery is not rare). Suppose, in a given species, the two relationships are equally frequent. The average degree of relationship is then $(0.5 + 0.25)/2 = 0.375$. A gene causing a nestling to behave in a way appropriate to this degree of relationship will spread: the nestling need not know the frequency of adultery, or that the coefficient of relationship for half-sibs is one-quarter.

Costs and benefits combine non-additively

The essential point is illustrated in Figure 5.2. The argument is given for groups of two, but can readily be extended to larger groups. It is assumed that the basic fitness, in the absence of co-operative acts, is two units. In the additive case, a co-operator gives up one unit of fitness, and benefits the other member of the pair by three units. Hence the fitness of each member of a pair of co-operators is $2 - 1 + 3 = 4$ units. In the synergistic case, it is supposed that, if both members of a pair co-operate, their fitness is greater than would be predicted from this additive assumption.

It is convenient to set out these fitness values in a pay-off matrix: in these matrices, the entries, or pay-offs, are the fitnesses of the individual adopting the strategy on the left, if its partner adopts the strategy above. In analysing such a matrix, we see an 'evolutionarily stable strategy' or ESS. This is a strategy that is uninvadable, in the sense that, if almost all members of a large population adopt it, no mutant strategy, different from the ESS, can do better.

[118]

Note that what is important is not which strategy does best in a mixed pair, but whether the mutant does better against the ESS than the ESS does against itself.

Consider first the additive case. This matrix is the same as that of the 'prisoner's dilemma' game. It contains an element of paradox. No matter what one's partner does, it pays to defect. Hence, if the players are rational, both will defect: yet both would be better off if both co-operated. In an evolutionary context, defect is the only ESS: as stated above, if fitnesses combine additively, and if assortment is random, it is only the cost that matters. It follows that selection acting at the level of individuals will favour behaviour which is not optimal for a group. If assortment was not random, and co-operators always assorted together, then co-operation would invade a population of defectors: in fact, from Hamilton's inequality, if pairs were always full sibs ($r = 0.5$), then co-operation would spread, because $rb > c$ (i.e., $0.5 \times 3 > 1$).

Now consider the synergistic case, with random assortment. Co-operation is now an ESS, because, if co-operation is typical, and defection a rare mutant, their fitnesses are 8 and 5, respectively, so defection cannot invade. But defection is also an ESS. Thus a population has two possible stable states: co-operation and defection. It will evolve to one or other of these strategies. If, as is plausible, to defect is the primitive state, how can co-operation be established? What is needed is that co-operation should rise to a critical frequency locally: it will then spread through the whole population. If neighbours are often relatives, this may well happen.

An example may make this argument more convincing. Lions live in social groups, of which the females form the permanent structure: females born in a pride remain in it, whereas males leave when they become adult. The breeding males in a pride have come from outside. They have sole sexual access to the females, driving off other males. Usually several males hold a pride, and all have sexual access to the females. They can be said to co-operate, in that they act jointly to drive off other groups of males, and they rarely fight among themselves over females. A long-term study of the lions of the Serengeti shows that the breeding success of a male increases with the number of other males in a group, at least up to a group size of four. Thus co-operation does have synergistic effects on fitness. However, it is common, though by no means universal, for the co-operating males to be related: i.e. to be themselves born in the same pride. This may have been important initially in raising the frequency of co-operation above the critical value. Now that co-operation is typical, it pays a male to co-operate with non-relatives, and about 40 per cent of groups contain unrelated males. This example may be typical, in that it involves both synergistic fitness effects and genetic relationship.

It is worth asking why the females live in groups. Packer and Pusey (pers. comm.) find that lions are at least as successful in hunting singly as in groups, and suggest that, for a common predator living in open country, group co-operation is needed to defend a kill against other lions, rather than to make the kill in the first place. If so, this is another example of the synergistic effect of co-operation.

Reciprocal altruism and the repeated prisoner's dilemma

If pay-offs are as in the prisoner's dilemma, is there any way in which co-operation between non-relatives can evolve? Axelrod has pointed out that co-operation can be evolutionarily stable if the game is played repeatedly against the same opponent. The argument is set out in Figure 5.3 for two strategies, defect, and tit for tat (TFT) – i.e. co-operate in the first game and in all subsequent games play as your opponent did in the last one. As the diagram shows, this converts the game into one in which co-operation (TFT) is an ESS (so is defect: there is again a problem in raising the frequency of TFT to a critical value).

This form of co-operation is very similar to 'reciprocal altruism', as proposed by Trivers. This requires that an animal should help only those that have in the past helped it, and therefore that it should be able to recognize other members of the group individually, and to behave differently towards them depending on how they have behaved in the past. There is evidence that some animals can achieve this. A striking example is afforded by the olive baboon. These animals have a gesture by which one individual can solicit help from another, usually against some third member of the group. If, on such occasions, the solicited animal does help, it gains nothing for itself, although it may benefit the soliciting animal. To help, therefore, pays only if it results in the helper receiving help on a later occasion. Packer ranked individuals according to two criteria: first, if solicited, did an animal help?; and second, if an animal solicited, did it receive help? He found a significant correlation between the two rankings, suggesting that help may indeed be reciprocated.

Baboons are among the more intelligent of social animals. Axelrod and Hamilton have pointed out that no high intelligence is required for the evolution of co-operation between sessile organisms. Thus, if an animal has only one neighbour, it does not have to learn to distinguish between individuals: indeed, even plants could evolve co-operation, given synergistic fitness interactions.

The social contract game

I now turn to a game that can be played only by men (Figure 5.4). Again, I suppose that two strategies are possible – co-operate and defect. The matrix 5.4(*a*) gives the pay-offs to an individual, 'ego', adopting the strategy on the

Figure 5.3. *The repeated 'prisoner's dilemma' game*

	C	D			TFT	D
C	4	1	10	TFT	40	19
C	5	2	GAMES	D	23	20

D is only ESS   TFT and D are both ESS's

Fig.5.3. The repeated 'prisoner dilemma' game.

Figure 5.4. *The social contract game*

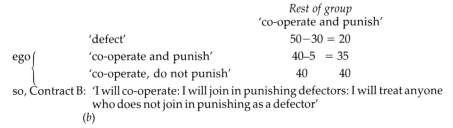

No contract

		Rest of group	
		C	*D*
ego	*C*	40	10
	D	50	20

If cost of being punished = −30
cost of punishing = −5

Contract A: 'I will co-operate: I will join in punishing defectors'

(*a*)

		Rest of group 'co-operate and punish'
ego	'defect'	50−30 = 20
	'co-operate and punish'	40–5 = 35
	'co-operate, do not punish'	40 40

so, Contract B: 'I will co-operate: I will join in punishing defectors: I will treat anyone who does not join in punishing as a defector'

(*b*)

Fig.5.4. The social contract game.

left, if the rest of the group adopts the strategy above. Suppose that the members of a group bind themselves to what I will call contract A: i.e., 'I will co-operate, I will join in punishing defectors'. The cost of being punished is −30, and of joining in punishing is −5. The latter is small because, in a group of reasonable size, the risk to any one individual in disciplining a defector would be small.

Unfortunately, contract A is not stable against what economists call a free rider: that is, an individual who co-operates, but does not join in punishing (Figure 5.4(*b*)). Stability requires contract B: 'I will co-operate: I will join in punishing defectors: I will treat anyone who does not join in punishing as a defector'.

I now discuss three questions:

(i) Is it true that non-human animals cannot play this game? If so, why?
(ii) Is there any related game that animals can play?
(iii) How far does the game mimic human society?

As an example, consider an agreement not to drive when drunk, and to punish those who do. Clearly, the concept 'driving when drunk' cannot be specified genetically. Some people might argue that no concept can be genetically specified, but I think this is a mistake. Curio found that hand-reared pied flycatchers will mob a model of a shrike (a natural enemy), provided the model is of the right size, colour and orientation, and has a dark horizontal line through the eye. This is clear evidence of a genetically specified concept, as, indeed, are all cases of innate releasing mechanisms identified by ethologists. The reason why 'driving when drunk' cannot be genetically specified is not

that no account can be so specified, but that driving has only existed for a few generations. It follows that an agreement to ban drunken driving requires communication in a language whose symbols have learned meanings. Other animals communicate symbolically, but (except, perhaps, for animals that have associated with man) it seems that the symbols used have genetically specified meanings.

A second reason why animals cannot play the social contract game concerns the way in which a contract arises in the first place. Some member of the group must, presumably, perceive that to ban drunken driving would be to his or her benefit, and set out to persuade others to join in the ban. To do this, the individual would have to appreciate that others might share his disapproval of drunken driving: otherwise, the attempt at persuasion would be pointless. This, in turn, requires that an animal should appreciate that other members of its species are individuals like itself, with similar likes and dislikes. I doubt whether animals other than man have this appreciation.

For these two reasons, and in particular because of the need for symbols with learned meanings, I think that a social contract can be achieved only by man. There is, however, something that animals may do that has many of the properties of a social contract. Suppose, in contract A, we replace 'defectors' by 'individuals that differ from the rest of the group'. If punishment consists of driving aberrant individuals out of the group, such behaviour could be favoured by natural selection: such individuals may well be suffering from infectious disease, or even be members of other, socially parasitic, species. There is much anecdotal evidence that animals that do differ, either naturally or because of experimental interference, are attacked and driven out: there is also evidence pointing in the opposite direction. I know of no critical review of these phenomena. In the present context, the relevant point is that the social norm to which individuals must conform need not be genetically specified.

A distinction must be drawn between the nature of the pay-offs in the social contract game, and in the games discussed earlier. In the earlier games, pay-offs were changes in Darwinian fitness. In the social contract game, they are the utilities of the various outcomes, as perceived by the players. Utilities must, I think, bear some relationship to changes in fitness, but it need not be a close one. The reason why there must be some relationship is as follows. At least some sensory inputs must be reinforcing or aversive without previous training: if there were no unconditioned reflexes, it would be impossible to acquire conditioned ones. Further, there must be some correlation between whether, on the one hand, an experience is reinforcing, and on the other, whether it increases fitness: if this were not so, learning would not increase fitness, and would not have evolved. However, the correlation between the pleasure or pain given by an experience, and its effect on fitness, will not be high, for two reasons. First, we live in an environment very different from that in which we evolved. Second, in a complex environment, very similar experiences may have very different effects on fitness. These points, of course apply to all animals, which as a result do things which manifestly do not increase their fitness: warblers feed baby cuckoos and moths fly into candles.

The social contract game was introduced to bring out some of the differences between human and animal societies. I now turn to two deficiencies of the game, considered as a model of human society. The first concerns the ways in which contracts arise and are enforced, and the second the complexities that arise because people do not all have the same set of options.

How are contracts enforced?

The contract model might suggest that social norms are reached by a process of reasoning and conscious acceptance, and enforced by means of law. In the case of a contract against drunken driving, this is roughly true. But for much of social behaviour, the processes are less explicit. The rules that must be obeyed are not written down, and may not be formulated in words, and enforcement is by means of social approval or disapproval rather than by legal sanctions. We are concerned with systems of custom and belief, taught by example, reinforced by myth and ritual, and maintained by social pressure.

Opinions will vary on how far human behaviour can be explained by conscious pursuit of perceived self-interest, and how far by adherence to social norms. In fact, neither account by itself would be adequate. Social norms arise because they serve the interests of particular groups. As a biologist, I am concerned with how human beings came to evolve the characteristics that enable them to be socialized. (I dislike this term, perhaps because I greatly disliked undergoing the process myself. I use it for want of a better. However, I mean it to cover, not only learning to wear the right clothes on the right occasions, but also the acquisition of values and beliefs.) Clearly, the nature of the customs and beliefs acquired is variable and non-genetic, but the capacity to acquire them does seem to be part of human nature. I can suggest three possible answers to the question of what selective forces led to the evolution of the human capacity to be socialized.

(i) The capacity was not selected for.

Gould and Lewontin point out that errors can result if we assume that some trait is present because it has itself been the target of natural selection. For example, the human chin did not evolve because it confers some advantage on its possessors: it arose because of interactions between growth fields in the skull and jaw, that changed for reasons that have nothing to do with the chin. Traits which are not themselves the targets of selection are likely to be features of human behaviour. No one supposes that the ability to solve differential equations evolved because, in itself, it increased Darwinian fitness: the ability is, presumably, the unselected consequence of mental capacities that did contribute to fitness.

Could the capacity to be socialized be such an unselected consequence? It is possible, but I think it unlikely, because, as I will explain, it is easy to think of a plausible selective explanation. (To argue that a trait has been a target of selection because one can think of a selective mechanism is not as unreasonable as it may sound. For example, no one supposes that the vertebrate eye is an

[123]

unselected consequence of something else, essentially because it is so obviously adapted for vision. Hence, if a trait does have an obvious function, that function is probably the evolutionary explanation of its presence.)

(ii) The capacity evolved because the group is the 'unit of evolution'.

A group whose members can be socialized is likely to be more successful, as measured by group survival, and by the production of new groups by splitting or budding. However, this will not necessarily lead to the evolution of the capacity. As explained on page 123, natural selection acting on individuals does not necessarily lead to the evolution of characteristics that are optimal for the group. If groups are to evolve characteristics optimal for their own survival and reproduction, they must be 'units of evolution', in the sense of having the properties of multiplication, heredity and variation needed for Darwinian evolution. Group heredity requires a high degree of reproductive isolation between groups. If such isolation existed, then groups might evolve traits ensuring their own survival. This is the process that Wynne-Edwards referred to as 'group selection': unfortunately, the term has recently been used in so many senses that it no longer has a clear meaning.

The snag with this explanation, then, is that it requires groups to be reproductively isolated. If, as is common in social mammals, the members of one sex (usually the males) leave the natal group and enter another before breeding, then the group no longer has heredity in the sense required for the evolution of group adaptations. Of course, we do not know what was the social structure during human evolution, but, by analogy with other mammals, it is unlikely that our ancestors lived in endogamous, inbreeding groups.

(iii) The capacity evolved because of synergistic effects on fitness.

Fortunately, we do not need to rely on group selection in Wynne-Edwards's sense. If a trait has synergistic effects on fitness, as defined on page 120, then it will be evolutionarily stable, although some degree of genetic relationship between group members will be necessary to establish the trait in the first place. It seems plausible that the capacity to be socialized would have strongly non-additive effects on fitness.

There remains, however, the question of why our ability to co-operate in groups relies so strongly on individuals acquiring shared values and customs, and a sense of loyalty to the group. It is possible to imagine a social animal in which co-operation really did depend on the rational acceptance of social contracts, as imagined in the model on page 121; indeed, such rational and conscious processes do play an important part in human co-operation. Clearly, it would be wrong to regard the two processes, of socialization and of rational agreement, as mutually exclusive. However, if individuals do acquire group values, and if they punish or ostracize those who do not act according to those values, this is a powerful way of coping with the problem of the free rider. It can be non-adaptive if circumstances change, leaving the group committed to beliefs that are no longer in the interests of its members, but in a stable environment it is highly effective.

Group identification in complex societies

The social contract game assumed that all individuals have the same actions open to them: that is, that they have the same strategy set. In complex societies, this is not true. The options are different for men who own factories or land and those who do not; for men with guns and men without; for men and women. Individuals belonging to these different categories have different options and different interests. To achieve their goals, they associate together in groups. Since these groups are more effective if they can command the loyalty of their members, they tend to acquire value systems of the kind discussed in the last section.

A crucial problem in understanding complex societies, therefore, concerns how individuals come to associate into groups. An individual will, of course, belong to more than one group: for example, a man might think of himself as a Welshman, a miner, and a freemason. However, different societies do seem to be dominated by different concepts of group membership. Marx argued that the predominant factor determining a man's social consciousness is his role in the process of production: is he a worker or a capitalist, a peasant or a landowner, a shopkeeper or a scientist? This approach to social analysis has been highly effective. However, there are societies – including capitalist societies – in which people's main loyalties are to racial, tribal or religious groups that cut across the economic class lines. In Ulster, people think of themselves as Catholics or Protestants – a distinction that has rather little to do with religious faith and a lot to do with history. I do not doubt that members of both groups would be better off if they forgot about these particular loyalties, but they show little sign of doing so,

For a biologist, it is natural to ask whether groups that command strong loyalties are necessarily 'breeding groups': i.e., are they groups for which children typically belong to the same group as their parents? Even if the answer to this question is yes – and I am not sure that it is – it would not be because groups are genetically different from one another: children are linked to their parents by cultural as well as genetic ties. Certainly, many powerful groups do have this property: for example, castes in India, blacks and whites in South Africa, Jews and Arabs in Israel, and, to a considerable extent, Catholics and Protestants in Ulster. Social classes in capitalist countries may approximate to breeding groups, provided social mobility is limited. There are two reasons why we might expect the groups that command the strongest loyalties to be breeding groups. Children are, to an important extent, socialized by their parents, and parents are ambitious for their children.

Although, as we shall see, there are exceptions, it is more likely that a group will become powerful if it is a breeding group. A realization of this fact seems occasionally to have led governments to take special measures to prevent a group, which might otherwise become a threat, from becoming a breeding group. Perhaps the most striking example was the employment of eunuchs to form the civil service in the Roman empire, a measure which may have helped to prevent the administrative class from acquiring excessive power: I do not know whether it was consciously planned with this end in view. A similar example is the formation by the Ottoman emperors of an army and palace guard, the janissaries, by recruiting Christian children, removed from their parents. Again. I do not know whether this was deliberately intended to avoid some of the risks of a military takeover. The role of celibacy in the Catholic church is more complex; however, it does seem to have made it easier for the church to act as an honest broker.

The political role of armies is an interesting one. Clearly, if an army can acquire the conviction that it is entitled to exercise political power, there is little that can stop it doing so. The weakness of armies stems from the fact that they are not breeding groups, and so do not easily acquire the necessary political conviction. In the past, however, armies have frequently evolved into feudal aristocracies, becoming breeding groups in the process. There are signs that the Communist bureaucracy of the Soviet Union may be undergoing an analogous transformation, although the economic structure of modern society is likely to make the end result very different.

Women constitute one group that cannot become a breeding group: half of a woman's parents, and half her children, are not group members. Yet women have manifest common interests. One could draw one of two lessons from the situation. On the one hand, one could argue that the inferior status of women in most societies, despite their numerical equality with men, confirms the fact that, to be really effective, a group must be a breeding group: if women had been able to develop a common ideology, their status would be better. Alternatively, one could point to the strength of the women's movement in contemporary western society, and to the ability of that movement to create its own mythology, counter to the mythology of a male-dominated society, as evidence that non-breeding groups can wield effective influence. I think that both views contain some truth.

At a more trivial level, the masons illustrate the potential power of a non-breeding group. Two points are worth making about them. First, I started this section by implying that groups would be constituted by individuals with a common strategy set: people with similar options open to them have common interests, and may therefore band together. But the masons demonstrate that the possession of a common strategy set is not a necessary condition for group formation. If the advantages of co-operation are synergistic, it can pay any group to co-operate. The second point is that, despite the obvious practical advantages that they derive from their membership, masons find it necessary to strengthen group loyalty by elaborate rituals and religious mumbo-jumbo.

A final question is as follows. How closely must the behaviour inculcated by a group actually lead to the furtherance of the interests of individual members? I think this question can only be answered in a historical context – i.e., in relation to the origin of groups, their maintenance, and their decline. In the early stages of group formation, the correspondence must be reasonably close. People with obvious common interests band together, and by degrees develop an ideology that strengthens and justifies the bond. One can see this happening at the present time in the women's movement, and in the gay liberation movement. However, as some of the examples mentioned above may suggest, there may be little correspondence, or even a contradiction, between the values of a group and the interests of its members. I would like to think that, when this happens, the group will in time disintegrate: one cannot fool all of the people all of the time.

In this last section, I have strayed a long way from evolutionary biology, the field in which I can claim some expertise. I would therefore not wish anything I have said to be taken too seriously, except as an indication of some of the problems that are worth detailed study.

Further reading

Axelrod, R. *The Evolution of Cooperation*. Basic Books, New York, 1984.

Dawkins, R. *The Selfish Gene*. Oxford University Press, 1976.

Maynard Smith, J. *Evolution and the Theory of Games*. Cambridge University Press, 1982.

Trivers, R.L. *Social Evolution*. Benjamin/Cummins, Menlo Park, California, 1985.

[6]

Origins of society

Ernest Gellner

I am honoured by and grateful to the Members of the Darwin College for inviting me to contribute to the 'Origins' series. I also forgive them the solecism which they have committed in so inviting me – mainly because I think they do not know that they have indeed committed a solecism. The subject which I am paid to teach in this University, Social Anthropology, has been defined for something like five or six decades by the prohibition, and that is not too strong a word, of speculation about origins. This is not just a precept built into the mores of the community which teaches the subject, but has in effect become part of its definition. I remember how, when I was becoming qualified in the subject by doing a Ph.D. on a North African topic, I once mentioned Ibn Khaldun, the very great fourteenth-century sociologist of that part of the world, to my supervisor. His reaction was to say: 'If you are going to talk about that, well, you can take yourself off and go into another subject.' So I did not mention Ibn Khaldun again until I actually had my Ph.D.

Now members of Darwin College would not attempt to bribe a policeman to let them park a car in some restricted street, because they know that they ought not to corrupt people. Nevertheless, this, in a sense, they have done to me, when I was offered an inducement to talk about origins. And the reason I am indeed talking about them is not because of the bribe, but because I think that prohibition of speculation about origins does call for serious discussion and revision. So I do have some serious scholarly reasons for doing that which, in a sense, according to the prevailing mores of my discipline, I am not meant to be doing at all.

When I was asked to give the lecture on which this chapter is based, I was asked to talk about the origin of *society*. When the programme actually appeared, I saw it was in the plural, and referred to *societies*. And this shift is entirely apposite: the relationship of the plural to the singular, of *societies* to *society*, provides one of the main and crucial clues for discussing the origins of society, the origins of human sociability as such, the origins of the kind of social order that we call a society. Of course, all this involves speculation. But such speculation ought not to be without interest, and it is precisely the relationship of the plural to the singular which enables one to make some advance in reformulating the question, and make it much more manageable. If you don't know the answer to a question, tinker with its formulation. That's a principle I learnt when I was a Professor of Philosophy.

The really crucial feature of what we call human society is its astonishing diversity. This diversity is not only very interesting in itself: it also provides the clue to the origin of what we call *society*.

The range of things which are called a human society is very extensive, and diverse societies do astonishingly different things. This is well known to be a problem, or to give rise to one, namely that of relativism. Given that there is diversity, and that sometimes diverse societies encounter each other or even live on the same territory, how do we know which is better, which embodies principles that should prevail? That is indeed a serious question, but it is not the one that concerns me here. The question is not: how do we cope with the consequences and implications of this diversity? but rather: how is that diversity possible at all?

People do not normally regard this as a problem. Diversification has obvious advantages. It gives one variety. The more options that are tried, the greater the chance of success. The fact that people are capable of constructing such very diverse social orders, and such very diverse cultures, has one immediate implication: a species capable of this diversity over space is also capable of diversity over time, and hence of sustained growth, whatever the desired direction might be. The specification of that proper direction is what bothers people who are bothered by the problem of relativism.

But the sheer *possibility* of diversity also constitutes a problem, and a very fundamental one, for the following obvious reason. Human beings are not genetically programmed to be members of this or that social order. You can take an infant human being and place it into any kind of culture, any kind of social order, and it will function acceptably. No doubt there are genetic constraints on what men can do. But these constraints are far wider than the constraints imposed by any one society. Each society narrows them down for itself, so to speak.

In other words, the diversity which is exemplified *between* cultures simply is not tolerable *within* any one of them. How is the unrestrained diversification of thought and conduct, of a cancerous growth in all the directions permitted genetically, inhibited in any particular society? This is the first question one has to ask about the origin of distinctively human societies.

How is this particular human kind of herd possible? Gregariousness as such is nothing distinctive; humanity has no monopoly on it, and gregariousness consequently doesn't present a specific problem. What does make human society so distinctive is not perhaps the existence of cultural diversification (there is some small measure of it in other kinds of species), but the truly fabulous range of it. The problem is; what *prevents* humans from developing too fast and too wildly, given that the genetic constraints are far too wide to explain the stability and homogeneity of human *societies*?

There is another human trait which I will discuss in an attempt to reformulate the question. In order to approach it, I will mention what is probably the most famous, but least tenable, theory of the origins of human society, namely that of social contract. This theory maintains that human society originates in a number of individuals who get together, and, for their mutual advantage, draw up an arrangement by which they thereafter abide, and that this is how society comes into being. I shall not discuss the theory at length, because it is well known to be absurd. The point I wish to make is that its absurdity contains lessons which have not been sufficiently exploited.

The objection to the social contract theory is of course that it is patently, brazenly, obviously circular. It presupposes the very thing which it is meant to explain, namely the existence of a contract-capable being, a being, that is to say, with the ability to conceptualize a situation distant in time and abstractly specified, and effectively to bind himself to behave in a certain kind of way if and when that situation arises. But the people who decry and ridicule the contract theory of society do not go on to exploit the illumination which this gives us as to what is distinctive about human societies. It is precisely the capacity to fulfil abstract obligations, and to conceptualize situations drawn from any one of a very wide and perhaps infinite range of situations. The number of situations which you can bind yourself to carry out is very wide, and presumably has that famous infinity which is attributed to language, and which constitutes the key premise for the Chomskian approach to language, and the rejection of behaviourism in linguistics.

How is a society established, and a series of societies diversified, whilst each of them is restrained from chaotically exploiting that wide diversity of possible human behaviour? A theory is available concerning how this may be done and it is one of the basic theories of social anthropology. The way in which you restrain people from doing a wide variety of things, not compatible with the social order of which they are members, is that you subject them to ritual. The process is simple: you make them dance round a totem pole until they are wild with excitement, and become jellies in the hysteria of collective frenzy; you enhance their emotional state by any device, by all the locally available audio-visual aids, drugs, dance, music and so on; and once they are really high, you stamp upon their minds the type of concept or notion to which they subsequently become enslaved. The idea is that the central feature of religion is ritual, and the central role of ritual is the endowment of individuals with compulsive concepts which simultaneously define their social and natural world and restrain and control their perceptions and comportment, in mutually reinforcing ways. These deeply internalized notions henceforth oblige them to act within the terms of prescribed limits. Each concept has a normative, binding content, as well as a kind of organizational descriptive content. The conceptual system maps out social order and required conduct, and inhibits inclinations to thought or conduct which would transgress its limits.

This is one of the central theories of anthropology, and until a better one is found, I shall remain inclined to believe that it must be valid. I can see no other explanation concerning how social and conceptual order and homogeneity are maintained *within* societies which, at the same time, are so astonishingly diverse as between each other. One species has somehow escaped the authority of nature; and is no longer genetically programmed to remain within a relatively narrow range of conduct. So it needs new constraints. The fantastic range of genetically possible conduct is constrained in any one particular herd, and obliged to respect socially marked bounds. This can only be achieved by means of conceptual constraint, and that in turn must somehow be instilled. Somehow, semantic, culturally transmitted limits are imposed on men.

The sheer diversity possible in a species of this kind, also makes change possible, change based not on any genetic transformation, but rather on cumulative development in a certain direction, consisting of a modification of the semantic rather than genetic system of constraints.

But this possibility of progress, which in our culture we think of as somehow glorious, presents a problem. Initially, the main difficulty facing societies was to *restrain* this excessive flexibility. The preservation of order is far more important for societies than the achievement of beneficial change, which only comes later, when conservation can be taken for granted, and when openings for genuinely beneficial change are available. Progress is possible because *change* is possible, because the internal construction of men allows such a wide range of conduct. But most change is not at all beneficial; most of it would disrupt a social order without any corresponding advantage. Before we can explain how beneficial change is possible, we must first show how too much change, with all kinds of chaotic effects, is avoided. This is not a Conservative Party political broadcast, but the point needs to be made. Conservation is the initial problem for a labile population, and it appears to be solved through deeply internalized concepts.

The question of course arises, which came first: this extraordinary ability and variety of possible conduct, or language? Obviously I do not know, but the point that has to be made is that the two are clearly correlative. Once you have the possibility of an astonishing range of behaviour, it has to be restricted somehow, and there has to be some kind of system of signs indicating the limits. This I assume is central to the origins of language. Language makes possible that astonishing kind of variety, by allowing it to be contained. It consists of a set of markers delineating the bounds of conduct, bounds whose *genetic* limits have become far too broad for any one social order. Only a species endowed with something like language can have a wide range of genetically possible behaviour; and a genetically broad range of conduct necessitating some mechanism – i.e., language – which restrains that which nature has failed to restrict.

Part of my point is that a number of things not normally connected must clearly have come together: the possibility of a wide range, the means of signalling the optional items within that range which come to be adopted by any one herd (in other words language), and the presence of ritual which socializes individuals into that system. Once you have this collection of traits, although you do not yet have the actuality of growth, you do have the potentiality of growth. So sociability proper, semantic rather than genetic transmission, language, diversity, ritual enforcement, and the potential for growth, make up a set of features which must all have arrived more or less jointly.

Our problem is: how do you generate an entire community which differs from an animal herd, not merely in the range of its potential cultural flexibility, but also in the capacity of its members to respect contracts? Ritual on its own does not yet give us the full answer: the kind of obligation instilled by ritual is not yet a contract. It is too rigid. It is unnegotiable, and it is not optional from the viewpoint of the individual who is subjected to it. It confers status on the individual subject to it, but does not yet enable him to subject himself to contract. A status is, if you like, a kind of frozen contract; and a contract is optional, choosable, a variable status. But a frozen status, like single-party elections, is a fraud, a travesty, not really a contract at all.

So one also needs to ask: by what conceivable steps can one proceed from a system of individuals grouped in social order, reasonably stable, no longer genetically constrained, to the kind of society that *we* consider normal, which we consider a real society, and which the authors of social contract theories consider normal? – a society in which men can commit themselves abstractly, at a distance, to a *kind* of behaviour, no longer frozen either genetically or ritually? Those who hit upon contract as a foundation of society gave a bad explanation, but in effect offered a good definition of our kind of society, of what it is that we are endeavouring to explain.

The ritually constrained system does not yet give us the kind of society that we can regard as normal and acceptable. A number of things had to happen after the emergence of a ritually, conceptually constrained system, before an order emerged in which people could optionally, rather than rigidly, concep-tualize and choose alternative patterns, and effectively bind themselves to abide by them. In other words, something further had to happen before the contract could become a model of social conduct. And once again, I think we should speculate concerning what those features were.

First of all there was the Neolithic Revolution, the development of a system for the production and storage of food. Human society was already highly diversified, even before this had occurred. The adoption of food production and storage was clearly within the potential of a concept-using and ritually restrained society, but it enormously expanded its possible size and com-plexity.

But it was also a tremendous trap. The main consequence of the adoption of food production and storage was the pervasiveness of political domination. A saying is attributed to the prophet Mohammed which affirms that subjection enters the house with the plough. This is profoundly true. The moment there is a surplus and storage, coercion becomes socially inevitable, having previously been but optional. A surplus has to be defended. It also has to be divided. No principle of division is either self-justifying or self-enforcing: it has to be enforced by some means and by someone.

This consideration, jointly with the simple principle of pre-emptive violence, which asserts that you should do unto them first that which they will do unto you if they get the chance, inescapably turns people into rivals. Though violence and coercion were not absent from pre-agrarian society, they were contingent. They were not, so to speak, necessarily built into it. But they *are* necessarily built into agrarian society, if by this one means a society possessed of a stored surplus, but not yet of the general principle of additional and sustained discoveries. The need for production and defence also impels agrarian society to value offspring which means that, for familiar Malthusian reasons, their populations frequently come close to the danger point.

The trouble with the popular image of Malthus is that it is a bit ethnocentric, and assumes a kind of stable institutional background, within which, if famine strikes, people die off at the margin: they wait patiently in a queue, and the people at the end of the line perish. But it does not work like that. The members of agrarian societies know the conditions they are in, and they do not wait for disaster to strike. They organize in such a way as to protect themselves, if possible, from being at the end of the queue. So, by and large, agrarian society is authoritarian and strongly domination-prone. It is made up of a system of protected, defended storehouses, with differential and protected access. Discipline is imposed, not so much by constant direct violence, but by enforced differential access to the storehouses. Coercion does not only underwrite the place in the queue: the threat of demotion, the hope of promotion, in the queue, also underwrite discipline. Hence coercion can generally be indirect. The naked sword is only used against those who defy the queue-masters altogether.

There are various kinds of situations where centralized domination does not occur, where some kind of balance of power engenders participatory and egalitarian polities. Nomads, some sedentary peasants in inaccessible terrain, and sometimes also trading cities, all exemplify this. But the overwhelming majority of agrarian societies are really systems of violently enforced surplus storage and surplus protection. These systems can vary in all sorts of ways, from collective tribal storehouses to governmentally controlled silos.

Political centralization generally, though not universally, follows surplus production and storage. It helps take us out of the first kind of social order, the system of ritually sanctioned roles, which might generically be called Durkheimian. Food production increases the size of societies: within large societies, the logic of rivalry and pre-emptive action generally leads to a concentration of power. A formalized machinery of enforcement supplements or partly replaces ritual. Food production and political centralization, and also one further very crucial step, jointly constitute a necessary rather than a sufficient condition of the next transformation, which leads towards the kind of society we are trying to explain. That additional third factor is the storage, not of a material surplus, but of meanings, of propositions, and of doctrine. This doctrinal and conceptual storage is made possible by *literacy*. This is that extra step. The feasibility of the storage and codification of ideas is as profound in its implications as the storage and socially enforced distribution of a surplus.

There is an asymmetry between the two storage discoveries. Material surplus generally, though not universally, makes for political centralization. And although political power and centralization in agrarian society is fragile, often unstable, it is nevertheless extremely pervasive. By contrast, the codification of concepts in writing does not lead quite as frequently to doctrinal centralization. Moreover, ideological doctrinal centralization has two aspects: organizational and ideological. The carriers of the codified origin may or may not themselves be centrally organized, and that organization may or may not have a single logical apex, a single guiding principle. The clerisy may or may not be united in a single organization with central leadership, and the ideas may or may not have an apex, so the two centralizations, ideal and organizational, do not necessarily go together.

The theory about what may then happen, and seems to have happened at least once, runs roughly as follows. If the doctrine is centralized and endowed with a single apex, for instance a single, exclusive, jealous deity, this may have very striking effects. The point about the ritually instilled concepts of preliterate, ideologically uncentralized, Durkheimian societies is that there is no need for them to be logically coherent. There is no earthly reason why the ritual accompanying the opening of the pasture should be in some way logically consistent with the ritual which accompanies weddings.

The point about the Durkheimian social order is that these concept clusters, these clusters of expectations and obligations, not only come in packages, but are also profoundly holistic in another way: they are not broken up into their elements. The various clusters do indeed tend to be articulated in the same style, thereby providing simple societies with that social coherence for which they are often envied. But they are not logically coherent, and indeed they are not expected to be.

I suspect there is a law which affirms that social and logical coherence are inversely related. From the kind of society endowed with a very high level of social coherence and logical incoherence, how does one reach the opposite condition, which appears to be ours, where a fairly high degree of logical coherence is accompanied by a minimal degree of social coherence? If the favoured explanation of the origins of logically diversified but socially harmonious conceptualization, is ritual, than the favoured explanation of the emergence of this other conceptual style is the impact of a rationalistic, centralizing, monotheistic and exclusive religion. It is important that it was hostile to manipulative magic, and insisted on a salvation through compliance with rules, rather than loyalty to a spiritual patronage network and payment of dues. It was a jealous Jehovah who taught mankind the Principle of excluded Middle. It is this that leads to the exclusion of that facile cohabitation of diverse conceptual schemes, of a logical tolerance which is so characteristic of simple societies, and also of what may be called non-Abrahamic traditions.

It is a plausible theory, and again, until we have a better one, I think it is the one which we must continue to work with. Of course, the jealous deity did not achieve all of this unaided. One may suspect the mysterious exclusivity of one high spirit had to have other support. It probably did have to be a high *spirit*. The person who noticed, and took at face value, the government of mankind by concepts, and built a theory around it, was Plato. Ironically, he did it at the very moment when locally and socially specific, ritual-born concepts were being replaced by script-born, potentially universal, trans-ethnic ideas. His Theory of Ideas confirmed the control-of-conduct through authoritarian concept-norms by the attribution of transcendent origin and authority to those norms. The Platonic theory of ideas is, all at once, a transcendence, and yet also a kind of coming unto self-awareness of the Durkheimian world: it recognizes that a society is a community of minds framed by *concepts*, which are, all at once, ways of clustering objects *and* ways of imposing obligations on men.

At the same time, he also sketched out what is really the generic social structure of agro-literate societies, namely government by warriors and clerics, by coercers and by scribes. In his version, the two ruling strata happen to be conflated, and top clerics were meritocratically selected from the authorized thug class. In historical practice, the details of this kind of government vary a great deal. Plato is relevant because he offered a marvellous blueprint of how this kind of society works and how it justifies itself.

His crucial mistake, however, was very much an intellectual one. He saw clearly enough that in this kind of world, the binding concepts are unequal, forming a kind of hierarchy, with a top concept ruling all the others. The Concept of the Good, as it were, generically incorporates and warrants the authority of all the others, which guide conduct at more specific levels. That theory may satisfy intellectuals, who are flattered to hear that their professional tools of trade, namely concepts, should be so sacred, and that the Top Concept should rule them all, and all of us. But most of mankind simply will not quake before a High Concept: they will, however, quake before a High God. This personalization was essential for the effective imposition of a centralized vision. To concentrate the mind of humanity, the apex of the system had to be *personal*, and the carrier of anger and wrath. But, at the same time, to achieve the effect which concerns us, it had to be a *Hidden* Deity; to set the rules and norms, but be too proud or too distant to interfere in day-to-day management of the world. It had to scorn making exceptions, it had to be distant and orderly, and could not be the kind of head of a bribable and interfering patronage network, which is what High Gods are in many other systems.

Once this notion of an orderly, distant, commanding, single apex prevails, certain other things became possible. The shift from ritual to doctrine, as the central agency of sanctioning the restraint which keeps society together, takes place, and this shift is supremely important. What goes with it is that the compulsion should no longer attach to individual concepts, but rather, to certain second-order features of conceptual living altogether. Concepts must be orderly and be applied in an orderly manner and be part of orderly systems. The orderly behaviour of concepts and of men is really attained by one and the same revolution.

Conceptually bound conduct, as stated, involves compulsion which is no longer genetically prescribed, but is instead culturally variable and ritually fixated. Concepts have a kind of grammar; at the place where one concept operates, another one could also be slotted in. Concepts are seldom *wholly* idiosyncratic; each is a kind of alternate within a system. But their grammar is initially rather rudimentary, in a Durkheimian world, because they don't break up into units, and shifting or replacing them is fairly restricted. It occurs only within limited sub-systems: there is no universal grammar, no incorporation of all ideas into one single system within one set of rules. But once sense of compulsion no longer attaches to individual concepts, but only to the whole orderly system, to orderly thinking as such, and to the orderly break-up of concepts into constituent elements, something else becomes possible: a society which is a system of contracts rather than a system of statuses, and a nature built up of evidence, and on the expectation of symmetrical order, and a society where both concepts and men are equal. Revelation, the unsymmetrical attribution of authority to some ideas and some men or some institutions, becomes impossible.

In one sense, *any* concept is a ritual: it is a named cluster of expectations and obligations, triggered off by socially prescribed conditions and contexts. What we have called a Durkheimian society lives in accordance with its pantheon of concepts, with their own hierarchy, and some are much more important than others, and are sustained by specially weighty rituals. The system as a whole is relatively fixed and non-negotiable. Now contrast this with a monotheistic, iconoclastic, puritanical, nomocratic world: a distant, hidden, rule-bound and rule-imposing, awe-inspiring God has proscribed magic, ritual, ecstasy, sacred objects, and enjoins a rule-bound morality on his creatures, and similarly, law-abiding regularity on all nature. He concentrates all sacredness in Himself, and piety is henceforth to be manifested in sober orderly conduct, in an undiscriminating observance of rules.

The sacred is now symmetrically diffused in the world, and no men are priests, special mediators, or rather, all men partake in priesthood equally. All men, *and* all concepts, are equal under God. Society is still concept-bound, but the concepts under which it lives and by which it is bound are equal, symmetrical and orderly. Reverence attaches to all of them equally; or rather, it attaches to their formal properties, to the fact that they form part of a unified and orderly system, and not to their specific and distinctive traits. None of them are specially underwritten by specially weighty rituals: the most potent ritual of the Protestants is absence of ritual, their graven image is the absence of graven images. All occasions in life now acquire equal weight, all affirmations are equally binding: no need for special oath-confirming rituals.

To put it pedantically, the authority of concepts has shifted to their formal, second-order characteristics. Now *Nature* in the modern sense becomes possible: it constitutes a unified law-bound system, within which no object, no event, no theory is sacred. There are no pre-empted, entrenched doctrines – which is the essence of theological religion. Equally, rational production (which is a precondition of capitalism, though not co-extensive with it) also becomes possible: it means the free, unrestricted choice of means in the pursuit of clearly specified and isolated aims.

A society makes the first shift from a religion centred on ritual and magic, and committed primarily to confirming and perpetuating stable status groups, when it acquires a class of literate scribes whose specialty is the codification of *doctrine*. In their competition with freelance shamans and possibly with other rival social groups, the scribes will stress the authority and primacy of doctrine, over which their literacy gives them a kind of monopoly. 'Reformations' are liable to be endemic in this kind of society: the scribes will seek to promote the exclusive authority of the doctrine as against rival forms of the sacred. Reformations, however, generally fail; but on one special occasion, a Reformation succeeded, at least in part, and a new kind of world emerged.

I am not putting forward an idealist theory of the emergence of modern society in terms of its ideology just because I have been stressing these aspects. I do not in fact hold any such theory. The elements which I have been stressing and trying to sketch out, are necessary but not sufficient; obviously a number of other things had to happen before a modern society could emerge. The separation of the clerical guild from the warrior stratum in the government of agrarian society, and a rivalry between them, and the occasional victory of the unified clerical organization was important; so was the neutralization of various groups of hereditary or professional coercers, effective political centralization, and the emergence of free production-orientated strata. All this made possible a simultaneous cognitive and productive explosion. One thing which held the status system rigidly together, apart from the old conceptual style, was also a system of rigid kinship; men had been caught between kings and cousins, and only escaped one of these to fall into the power of the other. At least two important theories have recently been put forward concerning the erosion of this constraint by Jack Goody and Alan Macfarlane. I do not know whether these theories are true, but they provide a model of how we could well have reached the point at which we find ourselves.

Once, however, social and economic conditions were favourable, this new thought style could emerge: it allows concepts to be freely dissociated and recombined, without sacralizing any of them. Constraint and compulsion are only imposed at a second level: men are obliged to think, and produce, in an orderly way. They are no longer tied to given specific concepts, to ways of looking at things, to given roles, to given production techniques. One of the great codifiers of the new vision was David Hume. His famous theory of causation says in effect that there are no inherent or given or obligatory clusters of things. Anything can go with anything, and the clusters can only be established by observation. First, all separables are to be separated, and then, their real associations established in the light of evidence, and of evidence alone. This is a recipe and a model for cognition, but it could equally of course be a recipe for market behaviour, the rules of which were plotted at the very same time by Hume's good friend Adam Smith. The combination and recombination of productive elements, unconstrained by conceptual clusters, by social roles in effect, is the essence of the liberal entrepreneurship, hallowed by Adam Smith. Its principles are the same as those which also underlie an unconstrained cognitive exploration of the world codified by Hume and by Kant.

The story offered here ends up with the kind of society which the contract theorists naively took for granted, when they invoked the contract as an explanation of that order which makes contract possible. The wide range of conceptualization, the free attachment of obligation at a temporal distance to any freely chosen content, is what we mean by contract. A society which has these features is also egalitarian, in the sense that it does not allow frozen systems of statuses; and it is also protestant in a generic sense, in as far as it does not allow a segregated species of privileged cognitive specialists. Individuals and concepts are equal, and form part of a system where the *form* may be hallowed, but the content is variable. This society is also *nationalist* in the sense that, for the first time in history, something happens which was inconceivable in the earlier period: the high culture, transmitted by writing and formal education comes to constitute the pervasive culture of the entire society, defines it, and becomes the object of loyalty.

What I have in effect done, as anyone familiar with this topic will have realized, is to give a very potted version of Durkheim and Weber, fused so as to form one continuous story. Durkheim's problem was, why are all men rational? By this he meant – why do men think in concepts, and why are they constrained by them? He understood the problem of conceptual thinking much better than empiricist philosophers, or empiricist anthropologists for that matter, such as Frazer. He understood that if concepts were formed simply by the process of learning from nature, by 'association', a semantic cancer would rapidly develop. There would be no limits on either behaviour or concept formation. If association engendered concepts, there would be no limit on the content of concepts, and the astonishingly disciplined behaviour of concepts and of men would be inexplicable. The fact that concepts mean the same thing to all members of a given community would become a mystery. So would the fact that they also impose discipline *on* men. This problem could not be solved in conventional ways, and Durkheim indicated the direction in which the solution lies.

Weber's question, of course, was not why are all men rational, but why are some more rational than others, where rationality is envisaged in marked contrast with Durkheimian rationality. It is the new and special rationality which concerned him, one that no longer quakes before ritually instilled concepts, but only respects certain formal rules concerning their deployment of all of them. Weber put forward a theory, which I have put before you in a simplified and exaggerated version, concerning how that new world came about, thereby making possible the cognitive and productive explosion.

He also made plain the cost of such a world. The price of the separation and levelling of all elements, the full utilization of the potential involved in uniting all concepts in a single orderly logical space, and obliging them to disassociate and re-associate at our convenience, is considerable. The price, of course, is the separation of fact and value, and the ending of that comfortable endorsement of social arrangements to which mankind had become habituated.

One hears many complaints about the absence of a good legitimation of our social order. The suggestion is that this really is a scandal, and that philosophy had better do something about it. If my account is correct, philosophy cannot possibly do anything about it. Admittedly, a large number of substitute re-legitimations or re-enchantments is on the market. They are not worth much.

What I have put forward is of course a theory; it is speculative. I have brazenly sinned against that famous injunction proscribing speculation about origins. The speculation, however, would seem to be more or less compatible with available facts; or at any rate, it is not blatantly in conflict with them. It explains them better than any available alternative, and it suggests further ethnographic, historical and other enquiries. As a good Popperian, I ask no more of theories. The irony of anthropology is that it was born of a passionate preoccupation with the question of origins. Somewhere between the forties and the eighties of the last century, eventually encouraged by Darwinism, this preoccupation engendered a distinctive discipline, which endeavoured to use contemporary simpler peoples as a surrogate time machine. That was what virtually defined the subject.

Then, in the 1920s, in the conceptual sterling zone, all this was overturned. It was overturned in an extremely fertile way, by Bronislav Malinowski. It is said that the greatest pleasures are overcome revulsions, and perhaps the most powerful taboos are overcome attractions. The new taboo on speculation concerning origins was indeed very powerfully internalized. Perhaps it is time to change once again. The point about pre-Malinowskian anthropology was that its data were not very good, but its questions were extremely interesting. Malinowskian anthropology is by now perhaps a bit the other way around. The data are admirable, the questions may be a bit stale. Perhaps the time has come to combine a high quality of data with a revitalization of questions. They may indeed be questions about origins.

Further reading

Burtt, E.A. *The Metaphysical Foundations of Modern Physical Science*. International Library of Psychology, Philosophy and Scientific Method, London, 1925.

Childe, V.G. *Man Makes Himself*. Library of Science and Culture, 5; London, 1936.

Durkheim, E. *The Elementary Forms of the Religious Life*. George Allen and Unwin, London, 1915; 1976.

Durkheim, E. *The Division of Labour in Society*. Free Press, Glencoe, Illinois, 1960.

Ellen, R. (ed.) *Between Two Worlds. The Polish Roots of B. Malinowksi*. Forthcoming: Cambridge University Press.

Firth, R. (ed.) *Man and Culture: An Evaluation of the Work of B. Malinowski*. Routledge and Kegan Paul, London, 1957.

Frazer, J.G. *The Golden Bough*. London, 1907.

Goody, J. (ed.) *Literacy in Traditional Societies*. Cambridge University Press, 1968.

Goody, J. *The Development of the Family and Marriage in Europe*. Cambridge University Press, 1983.

Goody, J. *The Logic of Writing and the Organization of Society*. Cambridge University Press, 1986.

Macfarlane, A. *The Origins of English Individualism: the family, property and social transition*. Basil Blackwell, Oxford, 1978.

Sahlins, M. *Stone Age Economics*. Tavistock, Chicago, 1972.

Weber, M. *General Economic History.* (Adelphi Econ. Ser.)
    London, 1927.
Weber, M. *The Protestant Ethic and the Spirit of Capitalism.*
    London, 1930 (reprinted George Allen and Unwin,
    London, 1965).

[7]

Origins of language

John Lyons

My title, it will be noted, is 'Origins of language', which might look straightforward enough. But it is not; and I am going to begin by indulging for a moment in that most favourite of academic pastimes, the drawing of terminological distinctions. What I hope to do is to demonstrate that what looks like a nice set of sensible, interesting and, at least in principle, answerable questions – how, when, where and perhaps also why did language originate? – turns out to be, on examination, one that is riddled with hidden presuppositions and ambiguities. All too often the questions that my title gives rise to have been discussed, even by specialists, without proper regard to the requisite distinctions. Consequently, much of what has been said and written about the origin or origins of language, not only by previous generations of scholars, but more recently, has to be interpreted with a considerable measure of caution. When we have clarified some of the questions, we shall be better able to decide which, if any, of them are answerable on present evidence.

I have just used the disjunctive phrase 'the origin or origins of language'; and in doing so, I have slipped in the first of my verbal distinctions, though it is not one that I shall spend a great deal of time discussing. The title that was given to me by the organizers of the series was the one that I have adopted: 'Origins of language'. I do not know whether they thought carefully before opting for this, in preference to, for example, 'The origin of language' (in the singular and with the definite article) or 'The origins of language' (in the plural, but also with the definite article). But it is interesting for a linguist, especially for one of a rather pedantic turn of mind like me, to observe that some of the lectures in this series bore the singular phrase in their advertized titles and others the plural. Whether or not there was any considered reason for this variety, it so happens that one of the live issues in contemporary discussion and speculation is whether language is of single or multiple origin. A point that will become clearer as we proceed is that there is both a more and a less obvious sense that we can, and should, associate with the phrases 'single origin' and 'multiple origin' – or, to use the more specialized jargon of the trade, with the terms 'monogenesis' and 'polygenesis'. I am more interested in the less obvious sense. I shall have little to say about the question whether all known languages have resulted from a single original language (at some remote period in the pre-history of mankind) or from several such original languages, each of them independent of the others. For the moment, let me say that I am quite happy that my title can be construed as implying that language is of multiple (and indefinite) origin, but in a rather different sense. (I will come back to this point.)

It has for some time been traditional for philologists and linguists speaking or writing on this topic to begin by referring to the relevant statute of the Société de Linguistique, founded in Paris in 1866: '*La Société n'admet aucune communication concernant, soit l'origine du langage, soit la création d'une langue universelle.*' ('The Society does not accept papers on either the origin of language or the invention of a universal language.') There are two reasons why I have felt that it is particularly appropriate, on the present occasion, that I should make this traditional and almost mandatory reference to the 1866 ban. The first is that I want to draw your attention explicitly to the date of the ban and to comment upon it in relation to the dates of some of Darwin's publications and those of his followers. The second is that, having explained why the ban was imposed in the first place and, more important, why the attitude that it betokens is still prevalent among linguists (whether they have ever heard of the ban or not), I intend to use the explanation as a peg upon which to hang the first of my terminological distinctions.

The first edition of *The Origin of Species* was published in 1859. *The Descent of Man* and *The Expression of the Emotions in Man and Animals* a dozen years or so afterwards, in 1871 and 1872, respectively. I will refer later to Darwin's own views on the origin (or origins) of language. As we shall see, although he did in fact use the singular and may well have subscribed to the thesis of monogenesis in the customary sense of the term, he explicitly defended the view that language is, in another sense, of multiple (or, at least, of dual) origin; and that is also the view to which I myself incline. Here, however, I want to emphasize the fact that it was at the very time when what was already coming to be known as Darwinism was being rapidly and enthusiastically adopted as the new scientific paradigm that the Linguistic Society of Paris, the most prestigious such learned society of the day, anathematized all speculation, whether Darwinian or not, about the origin of language. Why, it may be asked, did the founding fathers adopt such an apparently obscurantist attitude? Let me quote the great American linguist W.D. Whitney:

> No theme in linguistic science is more often and more volumin-ously treated than this, and by scholars of every grade and tendency; nor any, it may be added, with less profitable result in proportion to the labour expended; the greater part of what is said and written upon it is mere windy talk, the assertion of subjective views which commend themselves to no mind save the one that produces them, and which are apt to be offered with a confidence, and defended with a tenacity, that is in inverse ratio to their acceptableness. This has given the whole question a bad repute among sober-minded philologists. (1873: 279.)[1]

These words were written just a year or two after Darwin himself had written in *The Descent of Man*: 'I cannot but doubt that language owes its origin to the imitation and modification of various natural sounds, the voices of other animals, and man's own instinctive cries, aided by signs and gestures.' (1871.) Darwin's view of the origin of language combines elements of what have been nicknamed the *bow-wow* and the *pooh-pooh* theories, the former being based on the assumption that the first words were onomatopoeic and imitative of the barking of dogs and other animals, the latter on the assumption that they were derived from instinctive cries of pain and other feelings or emotions. These two theories were not of course original to Darwin. They had already had quite a long history and had long been in competition with other theories of the ultimate origin of language, some of which have acquired their own distinctive nicknames, such as the *ding-dong* theory and the *yo-he-yo* theory. I hasten to add, however, that these nicknames are not necessarily to be understood as scornful or dismissive. Nowadays at least, we can treat them as technical terms. The point being made here is simply that the theories to which they refer would have been condemned by Whitney and other 'sober-minded philologists' of the period as subjective and unscientific.

In fact, Whitney and his colleagues took the view that any enquiry into the ultimate origins of language was inevitably and irredeemably unscientific.[2] Either it was avowedly speculative and non-empirical (i.e., non-inductive) or, if it aspired to be inductive and scientific, it was based on evidence whose relevance or reliability could not be substantiated. The various theories that I have just mentioned fall into the first category: they rested, to quote a later writer, on the method 'of trying to picture oneself a speechless mankind and speculating on the way in which language could then have originated'. The alternative, empirical (or perhaps one should say, pseudo-empirical) approach to the question was usually based on the assumption that some languages were more highly evolved than others (richer in some sense or more complex), and that these more highly evolved languages had developed out of less highly evolved ones. Granted this undoubtedly plausible assumption, it remained, first of all, to find relevant examples of less highly evolved languages and, second, to show how they might have developed into more highly evolved systems. The problem was that by the late nineteenth century it had become clear to 'sober-minded philologists' (if I may be permitted to continue to use this phrase, and henceforth as a more or less technical term) that none of the very many languages of so-called primitive people, on the one hand, or languages of an earlier period, on the other, was in any relevant sense simpler, or less highly evolved, than any other.

For reasons that will occupy us in due course, there has recently been a certain renewal of interest in the question of the origin or origins of language. But so far, I think it is fair to say, this is confined to only a small minority of linguists. (I should explain at this point, perhaps, that, here and throughout, I am using the word 'linguist' to mean 'practitioner of linguistics'; and also that for present purposes it may be regarded as more or less synonymous with Whitney's 'philologist'.) The attitude of most linguists to evolutionary theories of the origin of language is probably still one of agnosticism.

Let me now introduce the first of my distinctions: between language in general and languages in particular. It was papers on the origin of language (*'l'origine du langage'*), rather than on the origin of languages (*'l'origine des langues'*), that were proscribed in 1866 by the Linguistic Society of Paris. This is an important point. One of the dominant concerns of nineteenth-century linguistics was to determine the origin of particular languages by grouping them into families and showing how the members of these families had developed, or might have developed, from either an attested or hypothesized proto-language: by showing, for example, how the Romance languages (French, Italian, Portuguese, Rumanian, Spanish, etc.) had developed from Latin; how the Germanic languages (English, Dutch, German, Swedish, Danish, Norwegian, etc.) had developed from Proto-Germanic and going back a stage or two, how Proto-Italic, Proto-Germanic, Proto-Celtic, Proto-Indo-Aryan, etc. developed from Proto-Indo-European. But surely, one might say, this implies an evolutionary theory of the development of language. And, so it does – in one sense. Indeed, it is compatible with, though it does not necessarily imply, a Darwinian theory of the origin of languages.[3]

The point I want to make, however, is this (and it will be clear how it relates to what I have just been saying about the date of the famous Linguistic Society ban and the reasons why it was imposed): the proto-languages of historical and comparative linguistics, whether attested like Latin or hypothetical, like Proto-Germanic or at one remove Proto-Indo-European, were assumed to be, by 'sober-minded philologists' of the late nineteenth century at least (if not by their predecessors), fully developed languages comparable in structure and function, in all relevant respects, with contemporarily attested languages. Comparative philologists of the early nineteenth century could be forgiven for thinking that, in reconstructing what we now call Proto-Indo-European on the basis of Latin, Greek and more particularly Sanskrit, they were reconstructing a language that was more similar to the so-called Ursprache, or original language, than any of the attested languages upon which the reconstruction was based were to one another. After all, they had no reason to believe that the length of time separating the Sanskrit of the Vedic hymns from the 'Ursprache' was appreciably greater than that which separates Vedic Sanskrit from present-day languages like German, French or English. The earliest Vedic records go back some 3500 years (to the middle of the second millennium BC): and, in the period to which I am referring, it was quite widely believed, both by those who accepted that the *Book of Genesis* gave a true account of the origin of the Universe, of man and of language and by those who did not, that the world and all that it contained had come into being no more than 7000 years ago. By the second half of the nineteenth century, however, the scholarly, if not yet the popular, view of the geological time-scale had changed dramatically. There was so far no directly interpretable evidence bearing upon the antiquity of either man or language. But it no longer seemed necessary, or even reasonable, to try to fit everything into a few thousand years.

Furthermore, partly for methodological reasons and partly as a consequence of sound empirical investigation, philologists and linguists of the late nineteenth century were beginning to formulate their own principle of uniformitarianism, comparable with that of contemporary geologists and palaeontologists. This can be interpreted either diachronically or synchronically. Interpreted diachronically, it implies that, however far back we go in time, in the historical description or reconstruction of languages, we shall find no evidence of any radical or essential difference between one kind of language-system and another. Interpreted synchronically (i.e., at any one point in time), but (if I may introduce the less familiar term) diatopically (i.e., in different places), it says that all the languages spoken at any one time, past or present, in so far as they are accessible to linguistic science, are very similar, if not identical, in structure and function. The principle of uniformitarianism also implies linguistic egalitarianism: the principle that all languages, in so far as they are of interest to the linguist qua linguist, are of equal worth.

Now the principles of linguistic uniformitarianism and linguistic egalitarianism have frequently been misunderstood by non-linguists. And they cannot be properly understood, I think, without discussing them at considerable length and bringing into that discussion far more of the conceptual apparatus of modern theoretical linguistics than is necessary or appropriate in the present context. Let me just recall that I introduced the principle of uniformitarianism by saying that it was based partly on methodological considerations and partly on the empirical findings of descriptive linguistics. Having reigned more or less unchallenged, within linguistics at least, for much of the twentieth century, it has recently been called into question, on both methodological and empirical grounds. However, I believe I am right in saying that none of the modifications or qualifications that informed opponents of it would wish to make are such that we need to take account of them here.

The relevance of the principle of uniformitarianism to our present concerns is that it rules out immediately two of the traditional empirical, or pseudo-empirical, approaches to the question of the origin of language. It prohibits the use of evidence from the historical, or diachronic, investigation of languages, on the one hand, and from the investigation of the languages of so-called primitive peoples, on the other. We may now look at the third of the empirical approaches: that of making use of evidence from the acquisition of language by children.

As soon as this is mentioned, there forces itself upon our attention an alternative interpretation of the phrase 'origins of language' to the one that I have tacitly adopted so far. Instead of asking the so-called phylogenetic question 'How and when did language originate in the remote history or pre-history of mankind?', we can ask the ontogenetic question 'How and when does language originate in the individual?'. And if we subscribe to a linguistic version of Haeckel's famous slogan, 'Ontogeny recapitulates phylogeny', we can make legitimate use of ontogenetic evidence in the formulation of phylogenetic hypotheses. (Interestingly enough, Haeckel's dictum, expressing what he saw as a fundamental biogenetic law operating in association with Darwin's principle of natural selection, was first formulated at almost exactly the same time as the Linguistic Society ban: in 1867. I mention this otherwise irrelevant fact, for the same reason that I mentioned the dates of Darwin's principal publications: to evoke the *Zeitgeist* and to contextualize the linguist's view of the origins of language in relation to Darwinian, and neo-Darwinian, evolutionism.[4]) But does the ontogeny of language recapitulate its phylogeny? Many researchers have argued or assumed that it does, in general if not in detail; and it is of course possible that the principle holds for language, even if it does not hold in respect of the origin and evolution of biological species. There is, however, no very convincing reason to believe that it does; and this must, at the very least, give us pause for thought.

We shall return presently, to the ontogenetic question. It so happens that linguists have generally found this more interesting, in recent years, than the perhaps empirically unanswerable phylogenetic question. But it is clearly the latter that the organizers of the Darwin lectures had in mind when they supplied me with my title. I will not therefore discuss the acquisition of language by children for its own sake, or for the light that it throws upon the child's mental and emotional development, although this is the context in which it has been discussed, and intensively investigated, by linguists and psychologists in the last 25 years or so. I do want to emphasize, however, that the ontogenesis of language is in its way no less mysterious – no less productive of that sense of awe and wonder which Aristotle saw as the mainspring of science – than is its phylogenesis. It is only because the one is a matter of mundane experience, something that we may have observed taking place in our younger siblings, our children and our grandchildren, whereas the other is assumed to have taken place in the remote past, and in conditions of which we have little or no relevant knowledge; that we find the one initially less puzzling than the other. We now know a lot more about the stages through which children pass in the acquisition of their native language than was known even a decade or so ago. But there is still no generally accepted theory of the evolution of language in the mind of the child.

Before we proceed, there is a further distinction to be drawn. This is the fairly obvious (once it has been explained), but none the less theoretically profound, distinction between language and speech. The fact that language is distinguishable from speech, not only in principle, but also in practice, is a matter of everyday experience for anyone who has been brought up in a modern literate society. Consider, for example, the activity that is involved in giving a lecture which, like the present lecture, is subsequently published in written form. The lecture is delivered orally by the speaker and interpreted, auditorily or aurally, by the audience. But what the speaker is saying – the lecture that he is giving – may have been written out in advance (as this one was), so that, as he reads it out, he is simultaneously transferring it from one medium to another: from the graphic to the phonetic medium. And if any of the audience wanted to transfer it back from its phonetic to its graphic realization, by writing it down word for word as they are listening or transcribing a tape-recording afterwards, they could do so. Similarly, the editor of the subsequently published lecture could take as his draft either the author's manuscript or a transcription of a tape-recording of the oral presentation; and, ideally, the two versions of this draft would be in all relevant respects identical. To generalize, then, the fact that we can speak what is written and write down what is spoken demonstrates the independence of language and speech. It demonstrates that, as far as their verbal component is concerned (and I will say something later about their non-verbal component), languages are independent of the physical, or psycho-physical, medium in which they are manifest: they have what I have elsewhere called the property of medium-transferability.

Indeed it is in my view demonstrable, though perhaps not so swiftly, that languages are totally abstract, non-physical, systems. Obviously, one's know-ledge of a language has a physical, and more precisely a neurophysiological, basis; and our now somewhat increased understanding of the neurophysio-logy of both speech and language is something that may well be relevant, as we shall see, to our present concerns. It is equally obvious that language – utterances, the products of our knowledge of this or that language (utterances that are identifiable as being in this or that language), cannot pass from speaker to hearer, from writer to reader, or more generally from sender to receiver, unless they are realized, or 'inscribed', in some physical medium and transmitted along some physical channel of communication. But the utterances themselves, like the languages of which, or in which, they are utterances have an abstract, non-physical structure, that gives each of them its identity and makes it possible for anyone who knows the language to classify them as what the semioticians call tokens of the same type. And, if one reflects upon it, one will see that this is what is implied by the fact that speech can be converted to writing, and conversely. It is, I repeat, a fact of profound theoretical significance.

It is nevertheless a fact that is frequently lost sight of in discussion of both the phylogenesis and the ontogenesis of language. All too often, language is identified with speech and evidence for the origin of the one is taken, uncritically, as evidence for the origin of the other. We must keep in mind at least the logical possibility of their separate and independent origin. And this is not just a logical possibility. One theory of the phylogenesis of language (to a modified version of which, as it happens, I myself incline) is that language did not originate in speech, but in gesture. This is not of course a new theory. Condillac, in his famous *Essai* of 1746, probably the most original and the most influential of the many eighteenth-century treatments of the origin or origins of language, argued for it. So too, subsequently, did Tylor, Morgan, Wallace, Wundt and many other scholars, representative of several different disci-plines. But now there is new evidence for the gestural theory. This comes partly from linguistics, but principally from physical anthropology, palaeon-tology, psychology, ethology and neurophysiology. I will come back to the gestural theory at the very end of the lecture.

If we think, for a moment, not about the when and the where of the origin of language, but rather about the whence and the how, we can postulate, in the abstract, two possibilities; and one of these, as we shall see, can be divided into two sub-possibilities. The first major possibility is that language comes from nowhere – or, perhaps I should say, from nothing: that its provenance is *ex nihilo* and *de novo*. (It sounds more respectable in Latin!) And one traditional answer to the *how*-question associated with the thesis of *ex nihilo* provenance, as far as the phylogenesis of language is concerned, is that of the *Book of Genesis* and of comparable works in other cultures: the hypothesis of divine creation. Now, I do not wish to defend the thesis of *ex nihilo* provenance, still less that of creationism. It is worth pointing out, however, not only that it is logically possible, but that it must not be rejected out of hand as scientifically disreputable. *Ex nihilo* provenance, if not creationism, has in fact been defended recently by scholars whose scientific and secularist credentials are beyond reproach.[5]

The alternative major possibility is that language develops out of something else. And it is this second possibility that splits into my two sub-possibilities, which I am going to relate to the classic geological and biological alternatives of gradualism and catastrophism. It will be easier to see what is involved, perhaps, if we first restrict our attention to the ontogenesis of language: the acquisition of language by children.

When I was talking about the attitude of our 'sober-minded philologists' of the late nineteenth century, I stressed the importance of distinguishing between language and languages, but I have made little overt use of this distinction since. Let me now re-introduce it, together with the distinction that we have just drawn between language and speech, in respect of the ontogenesis of language.

We commonly talk about the acquisition of language by children. We recognize, however, that, in any normal sense of the word 'language', it is impossible to acquire language, either language in general (whatever that might mean) or some structurally undifferentiated language, without acquiring a particular language (English, French, German, etc.) which is identifiable as such by virtue of its structural differences from other particular languages (Arabic, Chinese, Russian, etc.). It is of course conceivable that there is some 'inner' language of thought, innate or acquired, which is neutral between the particular 'external' languages, and that this 'inner' language – let us call it Mentalese – is such that its pre-existence is a condition of the acquisition of any particular 'external' language. This is a traditional rationalist view; and it has been revived, recently with support from Chomsky's theory of generative grammar.[6] It is further conceivable that language-utterances are translated out of and into Mentalese in the course of their production and interpretation. This too is a recognizably traditional view, and one which has recently been brought up to date by psychologists and psycholinguists influenced by Chomskyan generativsm. But the arguments in favour of the psychological reality of a universal language of thought which plays the kind of role that is attributed to it in the acquisition and psychological processing of diverse 'external' languages do not appear to me to be very convincing: and, in what follows, I shall discount the possible existence of this kind of 'Mentalese' – not, I hasten to add, as having been disproved or as self-evidently absurd (though some might indeed discount them on these grounds), but simply as not proven. I will take it as a valid assumption, then, that it is only when the child has acquired a language (*langue*) that we can say that he has acquired language (*langage*).[7]

And what about speech? As we have seen, language is independent of speech and does not presuppose its prior existence. The converse however, ontogenetically at least, is not true. The child cannot be said to have acquired speech until he has acquired language by acquiring a particular language; and in the normal course of events this will be what we call his native language. With these considerations in mind, we may now look briefly at the acquisition of language, which normally, but not necessarily, involves the acquisition of speech.[8]

All (physiologically and psychologically) normal children who acquire their native language in (socially) normal circumstances pass through a relatively constant developmental sequence, regardless of race and culture and regardless of the language that they are acquiring. The first stage, which begins some three months after birth, is that of babbling. The second, which starts towards the end of the first year is that of so called holophrastic speech, consisting of one-word utterances. This is followed, during the second year, by the third stage, characterized by the production of simple two-word and three-word utterances of a so-called telegraphic character. As the child passes through the second and third stage of language-acquisition, he gradually improves his control of the sound-system and may begin to make use of some of the grammatical distinctions of the language that he is acquiring. It will still be some time before we will be inclined to say that he has acquired the language (in fact, it is unclear that there is any definite terminal point for language-acquisition), but we might be prepared to say that he has by now acquired some kind of rudimentary language.

There are several points that I now want to make on the basis of this very brief outline of the first three stages of language acquisition. First, the child, in the normal environmental circumstances that I am invoking, is simultaneously acquiring both speech and language. And his progress from one stage to another appears to be relatively smooth and gradual. We can identify various stages, including the first three that I have mentioned, in his acquisition of language and speech. But the transition from each stage to its successor, though rapid, is not instantaneous. In fact, the boundary between one stage and another is somewhat indeterminate, so that it is impossible to say, at any particular moment, that the child has just passed from babbling to holophrasis or from holophrasis to so-called telegraphic speech. His progress, in short, is gradual, rather than catastrophic.

Second, everything characteristic of an earlier stage, except for babbling, continues to be part of the child's competence and performance, for a while at least, at the next stage. For example, the child does not suddenly, or even gradually, stop producing holophrastic utterances when he has reached the stage at which he is able to produce two-word and three-word utterances. What he does at the third stage, if I may so express it, is to produce a mixture of stage-2 and stage-3 utterances. And this is so at every subsequent stage in the acquisition of language and speech. Not only is the progress from one stage to the next gradual rather than catastrophic, but there is a sense in which the

[149]

earlier stage for some time overlaps with the later or is encapsulated in it. In fact, speaking very generally (and omitting certain qualifications that do not affect the import of the point I am making), we can say, of any stage after the second that it will encapsulate its predecessor, which in turn will encapsulate *its* predecessor, and so on. And this generalization holds (with certain qualifications) throughout the whole process of language-acquisition.

Third, the acquisition of language and speech is a natural process. In fact, it appears to be natural in at least two relevant senses of this rather troublesome word[9]. And it is important to distinguish them. First of all, it happens naturally in the sense that, unlike learning to read and write, it does not require instruction or training; and it is for this reason that the more general term 'acquisition' is nowadays preferred to the term 'learning', which has a more specific sense, both technically and non-technically, and might be held to imply the reciprocal term 'teaching'. One is not taught one's native language and arguably one does not learn it. One acquires it without formal instruction (i.e., naturally, in this first sense of 'naturally') as part and parcel of the process of growing up in normal environmental conditions, which include, of course 'exposure' to spoken utterances.

The second sense of 'natural' can be expressed by the phrase 'as a matter of biological endowment'. It has been argued, notably, in recent times, by Chomsky, that human languages are structurally and functionally adapted to the psychological nature of man: that they are, to use the biologist's term 'species-specific'. His argument that what he calls the language-faculty – the capacity to acquire language – is species-specific and genetically transmitted depends partly, but not wholly, on the fact that they are acquired naturally, in the first sense: without special instruction, as an integral component of the process of maturation and, it may be added (though Chomsky does not give much emphasis to this), socialization. There is much in the way that Chomsky develops this argument which is highly controversial and need not detain us in the present context. In particular, we need not be concerned whether the child's acquisition of language proceeds independently of his more general cognitive development and whether there is an innate language-faculty which is as highly determined by the principles of what Chomsky calls universal grammar as he supposes. It is perhaps worth noting, however, that, after a period in the 1970s when psychologists investigating language-acquisition reacted against the possibly excessive concern with grammatical structure characteristic of the 1960s and tended to favour cognitive and social explanations, they are now once again giving recognition to the child's interest in the purely formal patterns of grammar, which seem to be inexplicable from this point of view. Chomsky's hypothesis that the child is born with a language-acquisition device (LAD) of a problem-solving character, if not his notion of universal grammar, has been, to that degree at least, confirmed by recent work. As for the innateness of the child's predisposition to acquire speech (or, more precisely, to produce and recognize the vocal sounds and contours in which language is realized as speech): it has been demonstrated experimentally that from a very early age (long before he produces them) the infant is able to recognize speech-sounds, only some of which he can have heard in the spoken language to which he is 'exposed', and furthermore that, unless his ability to produce and perceive particular distinctions is reinforced by environmental 'exposure' it will, as it were, 'atrophy'. Let us grant, then, and although it may not be absolutely conclusive, there is a good deal of evidence for this, that human beings are genetically programmed to acquire both language and speech.

The question I now want to raise is whether, granted that language and speech are natural in this second sense, the link between them is also natural: in the second, rather than in the first or in perhaps some third looser sense of 'natural'. Most linguists, I think, assume that it is. But they may well be wrong. It is at least arguable, on present evidence, it seems to me, that the child is genetically programmed to vocalize and to produce and recognize speech-sounds, on the one hand, and to acquire a complex and flexible communication-system whose grammatical structure in particular is such that we would call it a language, on the other, but that the association of the two is a matter of environmental circumstances.

Evidence that the link between language and speech can be severed in abnormal environmental conditions comes primarily from recent work on the structure of gestural sign-languages used by the deaf. These can certainly be acquired naturally in the first of the two senses that I have identified. And, contrary to what used to be thought, there is no reason to deny that they are fully fledged languages or, alternatively, to say that they are parasitic upon or derived from spoken languages used in the same community.[10] The relevance of this point to the gestural theory of the ontogenesis of language is perhaps obvious. I will come back to it.

At the beginning of this lecture I referred to the recent renewal of interest in the phylogenesis of language; and I said that there was new evidence from a variety of disciplines. As far as linguistics is concerned, the renewal of interest dates from the 1960s and derives, interestingly enough, from two opposing viewpoints. One of these I have mentioned already: Chomskyan generativism, with its bias towards rationalism and its insistence upon the biological uniqueness of man. The other is characterized by its attachment to behaviourist psychology and to what one may in this Darwinian context call the ethological attitude.

A prominent representative of the second viewpoint was Hockett, whose main contribution to our now considerably increased understanding of the issues resides in his comparison of human language with animal communication systems in terms of selected key properties or design features. I do not propose to go into the details of Hockett's classificatory framework: I have done this elsewhere. Suffice it to say that it was of historic importance in giving an impetus to modern studies of animal communication systems and in focusing the attention of investigators on a common set of problems. But it suffers, I believe, from three fundamental flaws: first, it fails to distinguish language from speech and thus relates language more closely to vocal signalling, such as bird-song, than it does to non-vocal signalling, and, in particular, gestural signalling, among other primates; second, it treats all the properties as being simply present or absent, rather than as being present to different degrees; third, it treats as properties of languages as a whole those that are characteristic of only the verbal, in contrast with the non-verbal component of language.

There is no space to develop these criticisms here or to provide an alternative. Let me simply say that, if one takes account of the points that I have just made, one cannot but come to the conclusion, I think, that human languages have a component that is unique to them, both structurally and functionally, and another component that makes them comparable with the communication systems of other species.

Let me now refer briefly to the other kinds of evidence that have been brought to bear on the question of the phylogenesis of language in the last twenty years or so. It is impossible to deal with this in detail, since it is often highly specialized and can only be properly evaluated by someone who is competent in a large number of disciplines: psychology, physical anthropology, neurophysiology, ethology, palaeontology and many others. Furthermore, much of it is controversial. In fact, two of the reviewers of what is so far the most comprehensive, though perhaps not the most up-to-date, state-of-the-art report on research in this field, have said: 'the over-all impression is of controversy on every possible point'.

The relevant new evidence, such as it is, comes from the study of hominid fossils and making inferences about brain-size and the vocal tract; from the study of cerebral dominance and the lateralization of certain aspects of language-storage and language-processing in one hemisphere of the brain rather than the other; from the study of child-language, of sign-language and of creole and pidgin languages; from the study of animal communication systems and of non-communicative behaviour in respect of manual dexterity.

I will now comment, briefly and selectively, on some of the evidence to which I have just referred, in the light of the terminological points made earlier, beginning with what is *prima facie* most directly relevant to the origin or origins of human language: evidence that comes from the study of hominid fossils.

As I mentioned above, two topics are of principal concern here: first, the shape of the vocal tract; and, second, the size of the brain, in so far as this can be assessed accurately from the hominid skull fossils that have been investigated from this point of view. The relevance of the first of these two new research topics derives from: (a) the fact that the human vocal tract (the height of the larynx, the mobility of the tongue, the size and shape of the resonating chambers, the size, shape and disposition of the teeth, etc.) differs significantly from that of other contemporary primates; and (b) the assumption that the differences can be plausibly accounted for, in evolutionary terms, as the result of the adaptation of bodily organs whose primary biological function has to do with respiration and with the intake and mastication of food to the biologically secondary function of producing articulate sound which, when it serves as the medium for language, we call speech. The relevance of the second topic derives from the assumption that human languages, in the form in which we know them to-day at least, require a larger brain than is found in other contemporary primates, and perhaps also a brain in which the specifically human, and evolutionary 'new' or enlarged areas, are specialized for the storage and processing of language.[11] (It will be noted that I have deliberately related the first of the two topics to speech – and somewhat indirectly at that – and the second to language.)

Now, it is generally agreed that the development of the human vocal tract (in particular, the lowering of the larynx and the curvature of the supralaryngeal cavity) must be connected with the adoption of upright posture and bipedalism. This implies pre-speech (or, to be more precise, that the distinctively human vocalization of the kind that serves as a medium for language in human beings and, when it serves this function, is identified as speech) could have evolved from less distinctively human, non-linguistic or pre-linguistic vocalization with *Australopithecus* (i.e., some four million years ago) or with one of his successors in the line that leads ultimately to modern man: with *Homo habilis* or *Homo erectus* or even *Homo sapiens*. So far there appears to be no positive anatomical evidence that would justify the assignment of even an approximate starting-point for the evolution of the specifically human vocal tract or for the postulation of datable intermediate stages in its assumed evolutionary development.

There is of course the now famous, and controversial, reconstruction of a Neanderthal vocal tract and the computer modelling of the sounds that this would have been capable of producing. This purports to demonstrate that Neanderthals (despite their brain-size, which might suggest that they had the neural capacity both for the storage and processing of language and for the control of speech-production) were incapable of rapid and fully articulate speech. But two comments are in order here (apart from noting that the reconstruction itself is controversial): first of all, Neanderthal is commonly, if not universally, believed to be on a collateral line of descent from that of *Homo sapiens*; second, granted that Neanderthals of, say, 50000 years ago were incapable of rapid and fully articulate speech of the kind that is found in all known human societies today, it does not follow that they could not produce a subset of what we now identify as speech-sounds and at a slower rate.

Whether we opt for one date or period rather than another for the hypothetical starting-point of the evolution of speech or of its relatively rapid development from pre-speech into a more or less modern fully articulate system, making use of a wide range of sounds, depends partly, as we have observed, upon the assumption that speech and language evolved together *pari passu*. It also depends upon the correlation of such sparse and inconclusive evidence relating to the development of the vocal tract as is currently available with other kinds of evidence relating to mainly non-anatomical factors – ecological, climatic, cultural or social – and the association of these, more or less convincingly, with the adoption of bipedalism and with signs of intelligence or of the kind of symbolization that the possession and use of language is held to require, such as tool-making or the production of representational carvings and paintings.[12]

The fossil evidence relating to the size and configuration of the brain is also somewhat controversial and of itself equally inconclusive. It is generally, and no doubt plausibly, assumed that an increased brain-size/body-weight ratio is positively correlated, phylogenetically, with increased intelligence and with the evolution of language; and furthermore that the evolution of human intelligence and of human language are causally connected (possibly by means of a feedback relation). If we make these assumptions, we can interpret the evidence relating to brain-size as implying that none of man's ancestors earlier than *Homo erectus* could have had a more or less fully developed language. (A fully developed language, for present purposes, may be defined as one that comes as close to meeting the linguist's criteria of uniformity as all existing natural spoken languages do in terms of structural complexity and expressive power.) But, once again, it must be emphasized that the interpretation is based on far from self-evident assumptions, and also that the non-existence of what I am calling a fully developed language at a given period does not necessarily imply the non-existence of a less highly developed language at a considerably earlier period.

It is not just the fact of having a relatively large brain, however, that correlates with having a fully developed language in human beings. It is also the fact of having a brain with a particular cortical structure and with some degree of localization of specifically linguistic functions in certain areas of the neo-cortical tissue of the dominant hemisphere. I will return to this point in a moment. For the immediate purpose of evaluating the contribution that the fossil evidence has made, or might make in due course, to solving the problem of the origin, or origins, of language, all that can be said, I think, is the following: if we could interpret 'the patterns of the major gyri and sulci of the brain which can sometimes be seen in fossil endocasts' rather more confidently than appears to be possible at present, we would be on much safer ground than we are when we operate, perforce, solely with the criterion of brain-size. We might also be in a stronger position to decide whether the development, if not the origin, of both language and speech, is temporally and causally connected with 'sapientization'.

Let us turn to the question of cerebral dominance and lateralization, which has just been mentioned in connection with the interpretation of the fossil evidence and may now be considered in a more general context. The adult human brain is distinguished from that of other present-day primates, not only by its greater overall size (relative to body-weight,), but also by the greater development of the parietal regions, especially in the left hemisphere. Now, the left hemisphere is normally the dominant one as far as man's general cognitive abilities are concerned. Moreover, it contains Broca's Area which, since its discoverer Paul Broca (an early admirer of Darwin's *Origin of Species*[13]), started publishing his findings in 1861, has been popularly regarded as the 'speech centre' or 'language centre' in the brain. Not surprisingly therefore, it is generally, and once again no doubt plausibly, assumed that the development of the left hemisphere is causally connected with either the origin or the evolution of language. Unfortunately, as we have just seen, one cannot infer with certainty from the fossil record anything about the cortical structure of the brains of the ancestors of *Homo sapiens* or about cerebral dominance. What one can do, however, is to note one or two points about the lateralization of speech and language in present-day human beings which, on certain assumptions, can be seen as relevant to the questions that concern us.

The first point is that we now know that there is no single area of the brain in which language is stored or processed. Nor is it the case that all the processing of speech and language is carried out in the left hemisphere – or, indeed, by exclusively 'new-brain', neo-cortical, tissue: evidence from the study of certain kinds of aphasia suggests that the limbic brain has access to words associated with highly emotive associations. But there is still a very real sense in which the left hemisphere _ and, moreover, certain identifiable regions of the left hemisphere, of which Broca's Area is one and Wernicke's Area is another – can be correctly described as having a specifically linguistic function. As far as the reception of speech is concerned, it seems that what linguists refer to as the segmental phonemes (the vowels and consonants) are usually processed by the left hemisphere, but that the non-segmental features of spoken utterances (e.g., stress and intonation) can be handled equally well by either hemisphere. Both hemispheres are involved in the grammatical and semantic processing of language-utterances; but, once again, with some degree of specialization, the right hemisphere being able to interpret expressions referring to concrete objects, the left hemisphere alone being capable of interpreting more abstract expressions. Now, there are quite independent reasons, both structural and functional, for saying that the layers, or strands, in spoken language which are processed normally by the left hemisphere are just those that are uniquely or characteristically linguistic (i.e., unique to, or characteristic of, human speech and language, in contrast with animal communication-systems); and also that they are wholly contained within that part of language which is readily transferable from one medium to another.

The second point that I want to emphasize here is that the localization, or lateralization, of the more characteristically linguistic components of speech and language (under normal circumstances) in the left hemisphere is a maturational process that takes place over a period of the few years during which the child is (under normal circumstances) acquiring his native language. If for some reason (e.g., as a result of brain damage) lateralization is inhibited or one of the relevant areas in the left hemisphere is rendered inoperative, the other hemisphere can take over the characteristically linguistic functions – provided that the trauma or lesion that occasions the transfer of function occurs in early childhood. There is, then, a certain functional plasticity or flexibility as far as lateralization is concerned, but this is lost as the child approaches or passes what is often referred to as the 'critical age' for fully successful language-acquisition. And it is worth adding in this connection that there is some evidence to indicate that foreign languages learned after childhood may be handled by the non-dominant hemisphere.

What is the relevance of the two points that I have just made to the origin, or origins, of language? Taken together, they imply, to my mind, first, that not all the layers, or strands, in speech and language, have necessarily evolved at the

same time or at the same rate or in the same way; and, second, that the
localization of the more specifically linguistic functions in the dominant
hemisphere may be a relatively recent development on an evolutionary timescale. Admittedly, much of what I have been saying is somewhat speculative.
Also, the generalizations that I have been making are subject to innumerable
qualifications on points of detail; and I have strayed far from my own field of
expertise in making them. But they do contain a sufficient amount of what may
now be regarded as established fact. At the very least, they should disabuse us
of the all too common assumption that speech and language, in their entirety,
must have evolved together with cerebral dominance or the development of
neo-cortical tissue. So, too, should the discovery that, just as not all the
processing of language and speech is carried out by the 'new brain' in human
beings, so the communication systems of other primates may not be located
wholly in the limbic brain. The neurological evidence, as I understand it,
strongly supports the particular version of the hypothesis of polygenesis to
which, as I said earlier, I am myself attached (and for independent reasons): the
hypothesis that human language is a multi-layered or multi-stranded phenomenon, each of whose layers or strands may be of different antiquity and of
different origin.

I will not attempt to summarize the other evidence that has recently been
brought to bear on the question of the origin, or origins, of language.[14] Suffice it
to say that none of it, to the best of my knowledge, is inconsistent with this
particular interpretation of the hypothesis of polygenesis and some of it is
positively supportive. For example, it is now known that not all the phonetic
distinctions characteristic of human speech are innate and species-specific in
the strong sense of not needing to be learned by human beings and being
unlearnable by non-humans. Similarly, careful examination of the transition
from babbling to speech indicates that, even as far as segmental structure is
concerned, no sharp distinction can be drawn between pre-linguistic and
linguistic vocalization. In fact, much of the recent work on what has come to be
called pre-speech in children casts doubt on the possibility of establishing any
absolute discontinuities either between one stage and its successor in language-
acquisition or between human and non-human vocalization. In drawing
attention to research in child language-acquisition at this point, I am not of
course now adopting the thesis that ontogeny recapitulates phylogeny, which I
said earlier must be treated with caution. It suffices for the argument that I am
advancing here, not that the findings of recent research in child language-
acquisition should be held to give positive support to the hypothesis of
continuity in the phylogenetic evolution of language, but rather that they can
no longer be cited as readily as they have been in the past in favour of radical
and absolute discontinuity. It was to illustrate the difficulty of deciding
between catastrophism and gradualism that I referred earlier to the identifiable
stages of language-acquisition and to the fact that each stage overlaps with, and
is encapsulated in, its successor. We do not have to subscribe to a strong
version of the ontogeny-recapitulates-phylogeny thesis to consider that this
may also be the case as far as the phylogenetic development of language is

concerned. And it is nevertheless worth noting that the vocal tract of the human neonate is more similar to that of the adult non-human primate than it is to that of the adult human being and that some researchers have proposed developmental schedules which correlate identifiable stages in the earliest period of language-acquisition with the maturation of the more distinctively human anatomical and neurophysiological 'hardware' that is involved in the production and reception of spoken language.

Exactly the same conclusions that I have been drawing from the evidence of recent work in child language-acquisition can also be drawn, in my opinion, from the study of animal communication, and more particularly from the now famous chimpanzee experiments. Attempts have been made on several occasions in the past to train chimpanzees to use spoken language; and they have met with little success. More recently, three different series of experiments have been carried out with considerably greater success: one with a chimpanzee called Washoe, and subsequently with a second generation of chimpanzees interacting with Washoe; another with a chimpanzee named Sarah and a third with Lana. In none of the three cases, however, is it spoken language that the animals have been taught; for it is now recognized that the vocal apparatus of the higher non-human primates is not well adapted for the production of speech-sounds and this fact alone, it was thought, may account for past failures to teach them spoken language. I will not go into the details of these various experiments, which have been well publicized in the last few years. Suffice it to say that they were, in various respects, far more successful than most linguists, if not most psychologists and ethologists, would have anticipated. They have demonstrated that at least one species of non-human primates has the ability to acquire a communication-system with some degree of syntactic structure and productivity. Whether we say that their ability differs from the human capacity for language in degree or in kind is, in my view, largely a question of how we define 'language'. None of the various systems that the chimpanzees have learned has the grammatical complexity of the language-systems used by adult human beings. But they do not appear to differ significantly, in terms of formal complexity, from the language-systems of young children.

What is especially interesting, perhaps, is that some of the chimpanzees' utterances appear to be grammatically and semantically comparable with the utterances of children in what was described earlier as the third stage of language-acquisition. It has often been suggested that the utterances of children at this period can be accounted for partly in terms of expressive and social meaning and partly in terms of a small set of more specific structural meanings (vocative, desiderative, attributive, locative, agentive, etc.), such that the same combination of words may be associated with different structural meanings in different contexts. The chimpanzees' utterances, it has been claimed, can be analysed in terms of the same structural meanings and, considered in isolation from the context in which they occur, have the same kind of ambiguity or indeterminacy. Brown relates the set of structural meanings required for the analysis of children's two-word and three-word

[157]

utterances more particularly to the sensory-motor intelligence postulated by Piaget, with which, not only human beings, but also animals may operate and which develops in the infant, over many months, on the basis of his interaction with animate and inanimate entities in his environment. The implication is that the earliest, but not the later, stages of language development are under the control of sensory-motor intelligence; and that, as a consequence, we might expect certain species of animals to reach, but not go beyond, these earliest stages. In view of the structural and functional parallels that can be drawn between non-verbal communication (including the non-verbal component in language) and animal signalling-systems, one might perhaps go on to hypothesize that non-verbal communication in general is under the control of sensory-motor intelligence, whereas language in its fully developed form (though it continues to make use of the sensory-motor basis) requires a higher kind of cognitive ability. This hypothesis would also seem to be compatible with the facts summarized above relating to cerebral dominance and with what we know at present about the role played by the left and the right hemispheres of the brain in the storage and processing of language (and speech). However that may be, the fact that parallels can be drawn between the communicative behaviour of children and the communicative behaviour of chimpanzees casts doubt upon the view of those who would say that there is an unbridgeable gap between human and non-human communication.

I should perhaps mention at this point that many linguists, psychologists and philosophers would say that I have exaggerated the similarities between the communicative behaviour of young children and that of the chimpanzees that have been involved in the language-learning experiments. This may be so. In experiments of this kind, it is very difficult to control for the so-called 'clever Hans' effect, without introducing into the context in which the communicative behaviour is learned and exercised such a high degree of artificiality that comparability with the naturally acquired linguistic (and pre-linguistic) behaviour of human infants is seriously impaired. It is also very difficult, it should not be forgotten, to interpret the utterances of very young children in respect of intentionality and motivation.

There is the further problem that scholars working in different academic disciplines are subject to the influence of different professional prejudices. It is perhaps significant, as has been suggested, that 'the academic scepticism and even denigration which greeted Washoe's performance [and, subsequently, that of Sarah and Lana] . . . came principally from scholars working in theoretical linguistics and psycholinguistics – precisely those fields which might be said to have a vested professional interest in maintaining [a particular] concept of a language'. As a theoretical linguist myself, I feel in duty bound to compensate as best I can for the *déformation professionelle* to which I have, for many years now, been continuously exposed. At the very least, whilst recognizing that ethologists and animal psychologists might be said to have their own vested professional interest in seeking to establish the thesis of continuity and of the non-uniqueness of man, I have to concede that they have been right to complain about the arbitrariness of the linguist's decision that this or that set of 'design-features' is criterial for language 'properly so called' – and perhaps also about the linguist's tendency to move the goal-posts in the course of the match!

It is for much the same reason that I am inclined to subscribe, on present evidence, to the updated version of the gestural theory of the origin of language, to which reference was made earlier: because most linguists, committed as they are by the bias of their training to the primacy of speech, unhesitatingly reject it. The evidence in favour of the gestural theory is, admittedly, not very strong; and it has also been argued that the 'normal standards of parsimony' go against it. But the evidence, for and against, is so fragmentary and disparate that I am not sure that 'normal standards' of scientific parsimony are applicable. There are times, it seems to me, when one might just as well admit, to oneself and publicly, that one is inevitably going beyond the evidence and opting for or against a hypothesis, personally if not in the name of one's professional discipline, on non-scientific, but none the less intellectually respectable, grounds. For man does not live by scientifically justifiable hypotheses alone, but also by myth and more or less well-motivated prejudice. After all, *Credo quia improbabile* is logically and scientifically more defensible as a profession of faith than *Credo quia impossibile*! And all hypotheses in this area are, if not unprovable, so far unproven. Whether it is classified as myth or hypothesis, the gestural theory of the origin of language encourages the linguist to be critical of disciplinary orthodoxy and to think of language, independently of speech, in a broader theoretical context.

For what it is worth, the evidence that has been cited in support of the gestural theory comes not only from the chimpanzee experiments, but also from the study of primates in their natural habitat and from the investigation of sign languages (such as the one that was taught to Washoe); and this has been interpreted in the light of what is known (and has been referred to above) relating to early tool-use and symbolization, on the one hand, and to cerebral dominance in respect of speech and manual dexterity, on the other. As far as the relevance of sign-language is concerned, it is worth mentioning that some of the arguments that were advanced against it, when the gestural theory was revived, in its modern version, a decade or so ago, have proved to be fallacious. As we have seen, it is now known that the gestural languages of the deaf are neither parasitic upon co-existing spoken languages nor strikingly simpler than, or different from, other natural human languages in grammatical structure and expressive power – or, it would appear, in their degree of iconicity. Iconicity, of which onomatopoeia in spoken languages is the most obvious example, is generally regarded as one of the 'design-features' which separates non-linguistic, or pre-linguistic, systems of communication from fully fleged languages. But iconicity, more generally defined as non-arbitrariness of the association of form and meaning, is (like the other design-features to which I referred, in passing, above) not a matter of yes or no, but of more or less; and there is much more iconicity in 'ordinary' natural languages, at all levels of their structure, than the conventional wisdon in linguistics would have us believe.

On the basis of this and other evidence, including the fact that gesture continues to play an important 'paralinguistic' role in the modulation and punctuation of normal spoken utterances, it is argued that languages, as we now know them in their fully developed form, may have developed, whether by relatively slow evolution or catastrophically, between 100000 and 40000 years ago, not as a direct outgrowth of the expressive, or emotive, use of vocal signals characteristic of non-human primates, but out of a pre-existing system of manual gestures; and that this gestural system may have evolved at a time when man's hominid ancestors were adopting an upright posture, thus freeing the hands for this purpose and for tool-using, and when the brain was both increasing in size and acquiring the potential for the specialization of complex processing in the dominant hemisphere.

However, I would not wish to overplay my commitment to the theory, or myth, of the gestural origin of language. If it is correct, it implies that spoken language in more or less its present form has evolved by integrating structural and functional properties from different sources – some of which link us with other species and others which, arguably, make us unique. And regardless of the question of origins this is what I believe to be the case. Whether we emphasize our kinship with other primates, and more generally with other species, or our differences from them depends, as I have said, on wholly non-scientific considerations.

The reason why I spent so much time initially on clarifying the difference between 'language' and 'speech' and between 'language' and 'languages' (not to mention 'origin' and 'origins') will now be clear. The conclusions to which we have come, tentative and precarious though they may be, are, as I have said, either supported by or, at least, not contradicted by the findings of all the relevant disciplines, including linguistics. And possibly the most valuable contribution that a linguist can make to the discussion is to establish and promulgate the following proposition: the question whether language (i.e., human languages, as we know them) evolved from some non-verbal communication-system is not formulated precisely enough to be answered positively or negatively. And it must be emphasized that it is not just that we lack the evidence which would enable us to answer it. Although there is perhaps no sharp distinction between human and non-human communication and between language and non-language, there are certain properties of adult language at least, having to do with its grammatical complexity and its descriptive, or propositional, function, which appear to be unique to natural human languages (and of course to different kinds of artificial languages derived from them) and to be associated more particularly with their verbal component. If we decide to make the possession of these properties a defining characteristic of what we will call language, we can then say, correctly, that languages are fundamentally, or qualitatively, different from all other communication systems.[15] We might equally well have framed a definition of 'language', however, according to which we would be inclined to say that the difference between language and non-language is a matter of degree rather than kind. This purely definitional aspect of the question should be borne in